TRANSLOCAL GEOGRA

Translocal Geographies
Spaces, Places, Connections

Edited by

KATHERINE BRICKELL
University of London, UK

AYONA DATTA
London School of Economics, UK

Routledge
Taylor & Francis Group

LONDON AND NEW YORK

First published 2011 by Ashgate Publishing

2 Park Square, Milton Park, Abingdon, Oxon OX14 4RN
711 Third Avenue, New York, NY 10017, USA

Routledge is an imprint of the Taylor & Francis Group, an informa business

First issued in paperback 2016

British Library Cataloguing in Publication Data
Translocal geographies : spaces, places, connections.
   1. Acculturation--Case studies. 2. Emigration and
   immigration--Social aspects--Case studies. 3. Ethnic
   neighborhoods--Case studies. 4. Social ecology.
   I. Brickell, Katherine. II. Datta, Ayona.
   304.8-dc22

Library of Congress Cataloging-in-Publication Data
Brickell, Katherine.
   Translocal geographies : spaces, places, connections / by Katherine Brickell and Ayona
Datta.
      p. cm.
   Includes bibliographical references and index.
   ISBN 978-0-7546-7838-0 (hardback)
   1. Human geography--Cross-cultural studies. 2. Migration, Internal--Cross-cul-
tural studies. 3. Neighborhoods--Cross-cultural studies. I. Datta, Ayona. II. Title.

   GF41.B776 2010
   304.2'3--dc22
                                                                        2010036351

ISBN 978-0-7546-7838-0 (hbk)
ISBN 978-1-138-27269-9 (pbk)

Contents

List of Figures	*vii*
List of Contributors	*ix*
Acknowledgments	*xiii*

PART 1: INTRODUCTION: TRANSLOCAL GEOGRAPHIES

1   Introduction: Translocal Geographies   3
    Katherine Brickell and Ayona Datta

PART 2: TRANSLOCAL SPACES: HOME AND FAMILY

2   Translocal Geographies of 'Home' in Siem Reap, Cambodia   23
    Katherine Brickell

3   Translocal Family Relations amongst the Lahu in Northern Thailand   39
    Brian A.L. Tan and Brenda S.A. Yeoh

4   British Families Moving Home: Translocal Geographies of Return
    Migration from Singapore   55
    Madeleine E. Hatfield (née Dobson)

PART 3: TRANSLOCAL NEIGHBOURHOODS

5   Translocal Geographies of London: Belonging and 'Otherness'
    among Polish Migrants after 2004   73
    Ayona Datta

6   'You wouldn't know what's in there would you?'
    Homeliness and 'Foreign' Signs in Ashfield, Sydney   93
    Amanda Wise

7   Ways Out of Crisis in Buenos Aires: Translocal Landscapes and the
    Activation of Mobile Resources   109
    Ryan Centner

PART 4: URBAN TRANSLOCALITIES: SPACES, PLACES, CONNECTIONS

8   Fear of Small Distances: Home Associations in Douala,
     Dar es Salaam and London   127
     Ben Page

9   Translocal Spatial Geographies: Multi-sited Encounters of Greek
     Migrants in Athens, Berlin, and New York   145
     Anastasia Christou

10   Translocality in Washington, D.C. and Addis Ababa: Spaces and
     Linkages of the Ethiopian Diaspora in Two Capital Cities   163
     Elizabeth Chacko

PART 5: EPILOGUE

11   Translocality: A Critical Reflection   181
     Michael Peter Smith

Bibliography   *199*
Index   *223*

List of Figures

2.1	Channery's picture of her mobile phone	30
2.2	Yan's picture of his daughter on the Internet	33
2.3	Piseth's picture of his garden feature	34
2.4	Dith's picture of Notre Dame	37
3.1	Approximate Location of Research Areas	40
4.1	'My house in England' [Amy]	65
4.2	'This Buddah [sic] head is probably the most treasured possession from Singapore.' [Andrea]	67
5.1	Karol's picture of his car rear windscreen in Australia, in which he travelled across the country	80
5.2	Dawid's picture of his neighbourhood in East London	88
6.1	Typical Ashfield shops	95
6.2	The shop window referred to in the vignette below	101
7.1	madero½week at Opera Bay	119
7.2	La City, seen from Opera Bay (with World Bank in the crown of largest visible skyscraper)	120

List of Contributors

Katherine Brickell is British Academy Postdoctoral Fellow in the Department of Geography, Royal Holloway, University of London. She is currently working on a three-year project entitled 'Geographies of Transition in the Mekong Region: Gender, Labour and Domestic Life in Cambodia and Vietnam' (research which has been extended to Laos). Katherine has published her work in a number of international refereed journals including *Geoforum*; *Antipode*; *Signs: Journal of Women in Culture & Society*; *Progress in Development Studies*; and *Journal of Development Studies*.

Ryan Centner is Assistant Professor of Sociology at Tufts University. His research focuses primarily on Latin American cities and processes of socio-spatial change as they relate to citizenship struggles, planning interventions, and political-economic shifts. He is currently conducting a comparative project about the intersection of geopolitics, conflicts over local identity representation, and landscape redevelopment in Buenos Aires, Rio de Janeiro, and Istanbul. His work has appeared in the journals *City & Community*, *Local Environment*, *New Global Studies*, *Political Power & Social Theory*, and the *International Journal of Urban & Regional Research*.

Elizabeth Chacko is Associate Professor of Geography and International Affairs at the George Washington University in Washington, D.C. Her research focuses on migration, transnationalism and how immigrants modify and affect cities and spaces within them. She has worked on these aspects particularly with the Ethiopian and Asian Indian immigrant populations in the United States. Elizabeth has published her work in international journals such as *The Journal of Immigrant and Refugee Studies*, *Journal of Cultural Geography*, *GeoJournal*, *Openhouse International*, *The Geographical Review*, *Social Science & Medicine*, *Gender and Development* and *Health & Place*.

Anastasia Christou is Lecturer in Cultural Geography at the University of Sussex. She has been Research Fellow of the three year AHRC funded project 'Cultural Geographies of Counter-Diasporic Migration: The Second Generation Returns "Home"' under the Diasporas, Migration and Identities Programme (2007-2009) affiliated with the Sussex Centre for Migration Research. She has published widely on issues of migration and return migration; the second generation and ethnicity; space and place; transnationalism and identity; culture and memory; gender and feminism; home and belonging. Her recent book *Narratives of Place, Culture* *and Identity: Second-Generation Greek-Americans Return 'Home'* (Amsterdam University Press 2006) draws on all these issues.

Ayona Datta is Lecturer in the Cities Programme, Department of Sociology in the London School of Economics. Her current research project titled 'Home-building, migration and the city' explores how notions of home and the city are shaped through the experiences of East European construction workers arriving in London after EU expansion in 2004. She has published her work in a number of international refereed journals *Urban Geography*; *Gender, Place and Culture*; *Cultural Geographies*; *Environment and Planning A*; *Antipode*; and *Transactions of the Institute of British Geographers*.

Madeleine E. Hatfield (née Dobson), completed her PhD research at the Department of Geography, Royal Holloway, University of London and is Managing Editor: Journals at the Royal Geographical Society (with IBG). Her thesis is on the everyday experiences of return migration and homemaking, focusing on British households returning to the UK from Singapore and including children as well as adults. Madeleine has also researched the homemaking practices of students and the experiences of British skilled transients in Melbourne, Australia.

Ben Page teaches geography at University College London. He has worked and carried out research in Southwest and Northwest Cameroon for the last fifteen years. He is interested in environment and development relations, particularly in relation to the politics of water supply and in migration and development relations particularly in relation to the politics of belonging. This chapter draws heavily on collaborative work with Claire Mercer (London School of Economics) and Martin Evans (University of Chester). The research project on which this chapter draws was funded by the ESRC and more details can be found in the book *Development and the African Diaspora: Place and the Politics of Home* (Zed 2008).

Michael Peter Smith is Distinguished Professor in Community Studies and Development at the University of California, Davis. He has written extensively and published several influential books on cities and urbanism, global migration, and transnationalism. His scholarship on the interconnectedness of local and transnational social practices in specific cities is examined in his book *Transnational Urbanism: Locating Globalization* (Blackwell 2001). The politics of constructing and contesting 'extra-territorial citizenship' across the U.S.-Mexican border is a central focus of Smith's latest book, *Citizenship Across Borders* (Cornell Press 2008).

Brian A.L. Tan completed both his Master and Bachelor of Social Sciences in Geography at the National University of Singapore. His research interests include transnational migration, and Southeast Asian cultural and religious landscapes. He currently works at the Ministry of Manpower as a Senior Manager and has been instrumental in various foreign workforce policies in Singapore.

Amanda Wise is a Senior Research Fellow with the Centre for Research on Social Inclusion at Macquarie University, Australia. Amanda is an interdisciplinary researcher working at the intersection of sociology, cultural studies and urban anthropology. Her research into issues surrounding multiculturalism and migration draws primarily on ethnographic and qualitative cultural research methodologies. Her research interests include multiculturalism; racism and interethnic relations; diasporic, transnational and migrant communities; ethnicities; national and cultural identities; the senses, affect, and embodiment; forms of hope and belonging in urban Australia; cultural attachments and formations of place, hope and belonging in suburban Australia especially in relation to multicultural neighbourhoods; and developing theoretically and ethnographically informed anti-racism and 'social cohesion' interventions.

Brenda S.A. Yeoh is Professor (Provost's Chair), Department of Geography, as well as Dean of the Faculty of Arts and Social Sciences, National University of Singapore. She is also the Research Leader of the Asian Migration Cluster at the Asia Research Institute, NUS, and coordinates the Asian MetaCentre for Population and Sustainable Development Analysis. Her research interests include the politics of space in colonial and post-colonial cities; gender; migration and transnational communities.

Acknowledgments

We would first like to thank the Association of American Geographers (AAG) and the Royal Geographical Society (RGS) for sponsoring two sessions in 2007 and 2008 which really set off the ideas we developed in this book. We are immensely grateful to the Urban Geography Speciality Group in AAG which helped us bring together a group of speakers who remained interested in this project and contributed to this book after three years of gestation. We are also immensely grateful to the Developing Areas Research Group (DARG) and Population Geography Research Group in the RGS for sponsoring our session on 'Translocal geographies' where the initial idea of this book began to take shape.

The introduction and our individual chapters have benefitted from the innumerable discussions and questions we have faced from our audience in many invited lectures by Ayona at the University of Southampton, University of Essex, University of Sussex, and in a conference funded by the European Science Foundation in Sweden in August 2010 on 'Home, Migration and the City'. Katherine has presented initial work in 2009 as invited speaker to the Asia-Europe Foundation Alliance workshop in Bangkok, Thailand on 'Migration in the Southeast Asian Context', and at the 1st International Visual Methods Conference at the University of Leeds.

We would also like to thank Val Rose our editor at Ashgate, for her support of this project and the anonymous reviewer who gave us many helpful suggestions for improvement. Our appreciation also extends to Professor Katie Willis who provided comments on a number of the chapter drafts.

Finally Ayona would like to thank her partner and fellow migrant, Rohit Madan for always being there and sharing similar geographies of translocality.

PART 1
Introduction:
Translocal Geographies

Chapter 1

Introduction: Translocal Geographies

Katherine Brickell and Ayona Datta

Translocality draws attention to multiplying forms of mobility without losing sight of the importance of localities in peoples' lives (Oakes and Schein 2006: 1)

In recent years, there has been a coalescing of interest surrounding the notion of translocality, which increasingly appears in geographical work on transnationalism (see for example, Cartier 2001, Freitag and von Oppen 2010, Grillo and Ricco 2004, Castree 2004, Katz 2001, McFarlane 2009, Oakes and Schein 2006, Smart and Lin 2007, Tolia-Kelly 2008). Seen as a way of situating earlier deterritorialized notions of transnationalism which focussed largely on social networks and economic exchanges, translocality takes an 'agency oriented' approach to transnational migrant experiences. As Oakes and Schein (2006: 20) argue, translocality 'deliberately confuses the boundaries of the local in an effort to capture the increasingly complicated nature of spatial processes and identities, yet it insists on viewing such processes and identities as place-based rather than exclusively mobile, uprooted or "travelling"'. Taking into account the ever variegated localized contexts where transnational networks are maintained, negotiated and sustained in everyday urban life, scholars now assert the importance of local-local connections during transnational migration (Conradson and McKay 2007, Gielis 2009, Smith 2001, Smith and Guarnizo 1998, Mandaville 1999, McKay 2006a; Velayutham and Wise 2005). They suggest that localities need not necessarily be limited to the shared social relations of local histories, experiences and relations, but can connect to wider geographical histories and processes – in a way that articulates a 'global ethnography of place' (Burawoy 2000).

These calls to situatedness during mobility however, still retain national boundaries as the predominant focus of local-local connections. This is perhaps the history of translocality itself which has emerged from a concern over the disembedded understanding of transnational networks. Research on translocality primarily refers to how social relationships across locales shape transnational migrant networks, economic exchanges and diasporic space. In such an inquiry spatial registers of affiliation that are part of migrants' everyday embodied experiences remain largely unexplored. The effect of this has been to subsume the debates on translocality within a wider notion of transnationality. Translocality is now widely seen to be a form of 'grounded transnationalism' – a space where deterritorialized networks of transnational social relations take shape through migrant agencies. This means that translocality as a form of local-local relations exists primarily within the debates on transnationalism.

In this book we argue that there is a need to understand translocality in other spaces, places and scales beyond the national. Thus we are interested in translocal geographies as a simultaneous situatedness across different locales which provide ways of understanding the overlapping place-time(s) in migrants' everyday lives. As our starting point then, we understand translocality as 'groundedness' during movement, including those everyday movements that are not necessarily transnational. We call these translocal geographies because we take a view that these spaces and places need to be examined both through their situatedness and their connectedness to a variety of other locales. And in doing so, we open up ways of examining migration not only across other spaces and scales such as rural-urban, inter-urban and inter-regional but also bringing into view the movements of those supposedly 'immobile' groups who do not fall under the rubric of a transnational migrant but who negotiate different kinds of local-local journeys (both real and imagined). By grounding translocality within different scales and locales then, we are able to examine translocality beyond a notion of 'grounded' transnationalism (we will come back to this issue later in this chapter).

A theoretical and methodological challenge?

Such a conceptualization of translocal geographies provides us with both a theoretical and a methodological challenge. If we are to move beyond a focus on the primacy of national space, we have to map out how other spaces and places can become significant during the process of migration and movement. These may be interstitial spaces that are part of the itinerary of movement; they maybe sites from where movement and migration in other spaces and places are organized; or they may even be the corporeal body which moves across spaces. To theoretically account for these as constitutive of translocality means that we need to pay attention to their multiple and hybrid histories, their politics and social constructions, their material geographies, and their connections to other scales and places. If this means we pay attention not just to the transnational migrant; but also to those who move across other scales beyond the nation, we can no longer confine ourselves to the debates on transnationalism. Rather, we will need to take into account geographies of inter-regional or inter- and intra-urban movements. Finally, if we are to focus on the everyday materiality, corporeality and subjectivity of movement, we will also need to deploy a wider range of methodological tools which can capture not just the economic exchanges, political organizations or social networks across sites of departure and destination, but also the negotiation of wider range of spaces and place in between. Through this approach we believe we can examine the translocal geographies of everyday lives across spaces, places and scales.

Such a theoretical and methodological challenge in rethinking locality and its connections to other spaces is not without precedents. Over a decade has passed since Burawoy (2000) encountered a similar challenge in the articulation of a 'global ethnography' – one which extended the remit of ethnography from bounded insular research sites to analyses of the wider forces, connections and imaginations in a global world. The theoretical and methodological approach of global ethnography, he noted would allow for a more situated understanding of localized experiences 'as the product of flows of people, things, ideas, that is, the global connections between sites' (Burawoy 2000: 29). Burawoy did not lose sight of the importance of locality; rather he offered an understanding of different faces of globalization through its grounding in the local. Crucially, Burawoy highlighted the continuing relevance of national spaces in shaping the forces and connections in a global world.

Burawoy's insights are particularly significant for us since he highlights potential pitfalls of restricting ourselves to theoretical and methodological tools which are focussed primarily on micro-processes. In particular, his methodology of the 'extended case method'(2000: 27) which reaches out from micro-processes to macro forces, connects the local to the global without losing sight of the real experiences of globalization operating in particular localities. Yet we find that it is precisely in this significant connection between the global and local that 'global ethnography' misses a crucial opportunity – it does not provide a way to understand the local as situated within a network of spaces, places and scales where identities are negotiated and transformed. Instead of exploring the spaces and scales of the locale, it grounds the local simply as a site of negotiation of the global – a place where globalization is experienced by social actors.

How then do we conceptualize and do a translocal geography that rethinks the local through a variety of spaces, places and scales? What are the different theoretical and methodological tools through which we can map out how everyday spaces and places acquire meaning and saliency during migration? How do we simultaneously pay attention to the corporeality and materiality of movement and settlement, and the physical environments through which migrants traverse, without losing sight of how these are also shaped by their connections to other localities and scales of movement?

Space, places and connections are the three axes of the theoretical approach in this book. Existing literature on migration points to the centrality of space in a number of associated ways; in conceptualizing 'otherness' (Cohen 2004, Sandercock 1998), in anchoring memory and nostalgia (Blunt 2003) and/or 're-memory' (Tolia-Kelly 2004b). During migration and movement, spaces become simultaneously material and symbolic, located in the moral economies of the family (Velayutham and Wise 2005), in the material cultures of home (Tolia-Kelly 2004a, 2004b, Ureta 2007, Walsh 2006), in the socio-technologies of home-building (Datta 2008, Lothar and Mazzucato 2009), in urban neighbourhoods and 'monster houses' (Mitchell 2004). This book focuses on a variety of spaces which are related both to the local-local connections across transnational spaces, but also those that are part of more everyday mundane spaces of public transport, residential mobility, bodily and sensory perceptions that are negotiated during particular moments of migration and movement . We see these spaces as related to translocality at different scales that exist on a continuum from the corporeal body of the migrant to transnational spaces.

Reasserting this multiscalar approach to translocality means that we have to take seriously the material, embodied, and corporeal qualities of the local – the places where situatedness is experienced. And we have to examine the places where this 'local' resides. Of course, places are not inert recipients of migrants, but social actors are constituted by their interrelationships with, and their groundedness in, particular places. This is increasingly evident in the different encounters with 'others' that occur within particular physical locales, which are mediated through migrants' transnational histories, cosmopolitan attitudes, diasporic belonging, national identity, and particular positionalities of gender, race, ethnicity and citizenship (Silvey and Lawson 1999). For most migrants, a new physical environment implies new ways of interacting with people; it involves new kinds of behaviours in these places, new modes of movements, and new kinds of corporeal experiences. In all of this, the symbolic and material qualities of places are transformed. We take here a 'place-based' rather than 'place-bound' understanding (McKay 2006b: 201) of the local, which means that as people become more mobile, so too do locales become stretched and transformed (Castree 2004: 135). In other words, we take a translocal view of place, following on from Massey (1999: 22), that places 'may be imagined as particular articulations of social relations, including local relations "within" the place and those many connections which stretch way beyond it'. In her later thoughts on space, Massey (2005) takes this proposition further by suggesting that even as the political meaning of the local cannot be thought outside of contextual reference, it is also not enough to think of local/global linkages as an appropriate site of politics. Rather Massey notes that we need to take into account the political nature of places included therein and what kinds of interrelations between these places are produced from local/global linkages. This view of place raises the question – at what scale is the local constructed; at what scale is it politicized; and at what scale does it begin to have relevance in everyday lives?

During movement, spaces and places are invested with 'heightened material and conceptual significance' (Cairns 2004: 30), making them important bases for cultural understandings of relatedness (Datta 2009a, Datta 2009b, Datta and Brickell 2009). But as Wilding (2007: 337) notes, this does not mean an 'emphasis on processes, even as the multi-local "being there" is being asserted'. We examine translocal geographies as a set of dispersed connections across spaces, places and scales which become meaningful only in their corporeality, texture and materiality – as the physical and social conditions of particular constructions of the local, become significant sites of negotiations in migrants' everyday lives.

The methodological challenge of our task then has been addressed through an examination of diverse sites of translocality, which are multi-scalar. They focus on the personal experiences and narratives of migrants during their movements across nations, cities, neighbourhoods, homes and regions. Shaped variously by gender, class, age, ethnicity and migrant status, we see these stories of migration and movement as inherently spatial, linking different places at different times through a series of corporeal and subjective journeys. These journeys are also part of everyday life – across urban spaces, neighbourhoods, rental homes, shops, hotels, public transport and so on. Understanding these journeys involves not just detailed narratives, but also migrant biographies, participant observations, diaries and textual accounts. But these journeys are also hard to document simply through these tools. The authors in this book also utilized visual tools of videotaping, photo documentation and participant directed photography to bring in those 'absent' spaces and places in migrants' everyday journeys which are often hard for an ethnographer to access. Visual methods highlight the interconnected landscapes of migration, material embodiments of their journeys, and situatedness in particular spaces and places during everyday life. They move the discourses of migration as social networks, political organization or economic exchanges to multi-sited and multi-scalar translocal geographies.

This diversity of textual and visual methodological tools provides us with greater interpretive possibilities, nuance, and coherence to the range of embodied movements which are not captured within the transnational lens. They allow us to focus especially on the politics of subjectivity, corporeality and materiality during movement, which go beyond traditional approaches of settlement geographies and political economies. It is these variegated methodological tools which make visible a range of spaces and places that are part of migrants' diverse registers of affiliation. When applied to a wide range of case studies from Europe, Australasia, Southeast Asia, Africa, North and Latin America, they suggest that localities are still crucial to understanding experiences of migration and movement. But these localities are also constructed through their reference to other scales and to real and imagined boundaries and territories. Using these methodological tools, we can look in-depth at translocality as a situated mode of human agency and mobility through variegated spaces and places across nations, regions, cities, neighbourhoods, buildings and bodies. These are part of the everyday politics of agency during movement that engages with locally specific configurations of identity, (non) belonging and spatial practice; but which at the same time bring to the fore a variety of situated frames of reference beyond the nation.

In the next few pages, we turn to the notion of a 'grounded transnationalism', which has much to contribute to translocal geographies. We take grounded transnationalism as our point of entry to situating movement but also our point of departure for understanding the varieties of other spaces beyond the nation. We then look at how these other spaces can be situated across scales through a notion of habitus that takes translocality as a 'field' of everyday practice across scales. Finally, we outline the diverse registers of affiliation during migration that produces a multi-sited and multi-scalar translocal geography.

'Grounded' Transnationalism

For many national citizens, the practicalities of residence and the ideologies of home, soil and roots are often disjunct, so that the territorial referents of civic loyalty are increasingly divided for many persons among different spatial

horizons: work loyalties, residential loyalties, and religious loyalties may create
disjunct registers of affiliation. This is true whether migration of populations is
across small or large distances and whether or not these movements link traverse
to international boundaries. (Appadurai 1996b: 47)

Appadurai was writing about the translocal at a particular historic moment
when anthropological imaginations of the local were increasingly seen as limiting,
sedentary and insular. Critiquing the implicit acceptance of territorially bounded
nation states as the regulator of locality, Appadurai (1996a, 1996b) asserted that
contemporary local life is divided along a range of spatial horizons that frequently
go across, beyond and often without reference to the borders, and imaginaries of the
nation. Appadurai (1990, 1993) defined nationalism less by territorial sovereignty,
and more by the multiplicity of mobile practices enacted among refugees, tourists,
guest workers, transnational intellectuals, scientists, and undocumented migrants
whose lives are experienced through identities and aspirations which are not always
rooted in, or to, national coordinates. This was a view shared particularly in the case
of global elites who were held as the archetypal exemplars of disembeddedness
accompanying the shift in relationships between mobility, territory and national
affiliation (Hannerz 1996).

At around the same time when Appadurai (1996a; 1996b) was calling for a
deterritorialization of connections through translocality, Mitchell (1997) was
calling for a geographic imagination to be reinserted within research and writing
on transnationalism. Often seen in opposition to globalization, transnationalism
had examined migration beyond economic and demographic perspectives to
an understanding of migrant experiences and lived spaces, and in so doing had
articulated the different social, cultural, political and economic networks of migrants
across national borders. Mitchell was concerned about a sense of dislocatedness
that had come about in these articulations of transnational connections. For
Mitchell, a 'grounded' sense of transnationalism was important to examine the
spatio-temporal constructions of migrants' experiences and the ways that spaces
and places were not simply backgrounds, but played active roles in the dynamics
of mobility and movement. This grounded transnationalism is our point of entry
into translocality.

The paradoxical call for simultaneous situatedness and deterritorialization
in the late 1990s, can perhaps be understood better as the complex historical
trajectories of these two concepts. In the 1990s, transnationalism as the notion
of increased movements and connections across national borders focussed a
large part of its attention on the making and managing of global networks of
economic exchanges and entrepreneurship (Hannerz 1997; Skrbis 1999; Smith
and Guarnizo 1998; Werbner 1999). These networks were examined for the ability
of migrants to make and manage diasporas, business connections, political links,
and livelihoods across large physical distances. Translocality on the other hand
emerged from anthropological challenges to the investigation of bounded locales.
It attempted in the 1990s to explode a notion of insularity invested in the locale

to look at transnational connections without the need for situatedness. But, in the new millennium, as transnationalism became increasingly grounded, translocality came increasingly to be seen as an aspect of transnationalism.

There were many reasons why this came about. With calls for a grounded transnationalism, it became apparent that many of the transnational networks are shaped through the specificity of locales. In a challenge to the deterritorialized conceptualizations of translocality, Werbner (1997: 12) noted that transnational networks are based on loyalties that are 'anchored in translocal social networks rather than the global ecumene'. Transnational migrant entrepreneurship connected uneven geographic development to the trajectories of migrants from the developing to the developed nations, their economic activities, and their remittances sent back to particular towns and villages in their homelands (Olson and Silvey 2006, Harney 2007). Transnational missionary work within specific local contexts created religious communities across national borders (Mahler and Hansing 2005). And the transnational movement of aid workers in development organizations and aid agencies were facilitated by localized development practices (Bebbington and Kothari 2006). Transnationalism therefore began to be seen as a multifaceted, multi-local process in which diasporic communities were simultaneously emplaced *and* mobile – producing what Katz (2001: 724) called a 'rooted transnationalism'.

Grounded or rooted transnationalism in many ways has shaped the trajectory of translocality. Even as transnationalism was being called to become more situated, it was becoming clear to scholars that these transnational connections were only possible through local-local connections across national spaces. Writing on the relations between a town in the Dominican Republic and that of a neighbourhood in Boston, Massachusetts, Levitt (2001: 11) contended that people 'are all firmly rooted in a particular place and time, though their daily lives often depend on people, money, ideas, and resources located in another setting'. Through the political participation of Mexican migrants in California and Mexico, Smith (2003) highlighted similarly that the practices of transnationalism, while linking collectivities across more than one nation, are actually embodied in relations which are situated in specific local contexts (Smith and Guarnizo 1998). Transnational migrants were therefore never bereaved of locatedness; rather they were always socially and spatially situated actors rooting the transnational in the place-making practices of the translocal (Smith 2005a).

Grounding transnationalism within specific locales also meant that scholars could move away from examining migrant subjectivities as overwhelmingly linked to structural limitations, such as in the case of asylum seekers or refugees (Koser 2007) and focus instead on social agencies of migrants in everyday spaces. They showed that these 'coping strategies' of poor migrants were not restricted to their economic livelihoods; rather to the 'repertoire and mobilisation of skills and expertise that require the forging of noneconomic, social, and cultural allegiances' (Kothari 2008: 501-502). While transnationalism research had focussed earlier on 'juridical-legislative systems, bureaucratic apparatuses, economic entities, modes of governmentality, and war-making capabilities' (Ong 1999: 15), a grounded notion of transnationalism which emerged from these critiques, brought the importance of locality to the forefront.

Research on transnationalism became more nuanced and sophisticated through attention to its situatedness, by articulating a notion of translocality that was based on local-local connections across national boundaries. Translocal then became situated within the transnational – as a local site of exchange made possible through the movement of people or ideas across national spaces. In turn, translocality became a form of grassroots transnationalism – the face of 'transnationalism from below' (Smith and Guarnizo 1998) – as a process and practice linking multiple locations via local resistances of the informal economy, grass roots activism and the border crossings of unskilled migrant workers, refugees and exiles. It became a 'subset' of transnationalism, since it was believed that while being able to encompass many of the broader relations reflected by transnationalism, the term translocal also allowed an additional emphasis on place as a process of actual everyday relations (Velayutham and Wise 2005).

But, even as translocality focuses on local-local connections among transnational migrants, there has been relatively less theorization of movement within national boundaries or across localities that are situated within the nation. A number of scholars (Skeldon 2003, Deshingkar 2005) argue, for example, that the recent focus on international migration has drawn the attention of research away from internal movements of people. This potentially obscures the fact that internal migration is often far more important in terms of the numbers of people involved (see 2009 Human Development Report entitled *Overcoming Barriers: Human Mobility and Development*). These migrants are not transnational; rather translocal, but they also move across different places and equally take a range of decisions around moving. But these migrants often fall under the rubric of other forms of migration such as rural-urban or regional migration, which are often located within the discipline of development geography.

This then is our point of departure. In this book we retrieve translocality from within the confines of transnationalism to examine local-local connections in their own right and without privileging the national. We suggest that the nation-state is less than a potent vessel for understanding migrant identifications, and therefore the subsuming of translocality within transnationalism implies 'that the only things we are interested in are the dynamics across or beyond nation states, or within the (nation-) state system' (Levitt and Khagram 2004: 5). Translocal geographies, as we examine in this book then, includes migration in all its forms; it includes highly mobile and elite transnationals as well as those who are 'immobile' and often viewed as parochial; and it includes a focus on local-local movements that are part of a continuum of spaces and places related to migration.

There is another significant consequence of examining translocality in its own right. It allows us to examine the local as situated across a variety of scales – body, home, urban, regional or national; which means that translocal geographies can becomes a set of local-local negotiations across these different scales. This resonates with Ley's suggestion that, 'There is a need to re-incorporate other scales, including the regional, the national and the supranational but not yet global, such as the European Union and other continental-scale trading networks. ... Moreover, jumping scales is an important necessity both in explanation and in political practice' (Ley 2004: 155). In order to locate translocal geographies across scales, 'metaphors of domination need to be mingled with metaphors of vulnerability, images of global reach with those of parochialism, a discourse of detachment with one of partisanship' (Ley 2005: 157). Taking into account such diversity of scales and the spaces therein, we can examine migrants' locations within simultaneous positions of power and powerlessness in different places; migrants' access to different spaces of social networks and capital; and migrants' embodied and material locations across different places that are not necessarily about transnationalism but rather about translocal geographies of movement.

Situatedness across scales

How then do we theoretically examine the multiscalar situatedness of migrants' experiences? How do we account for the different ways that their experiences reach across different scales of movement – regional, national, rural, urban, neighbourhood, homes and bodies? How do we examine these scales not as separate entities; rather as a set of multiple affiliations enacted across space and time(s) by those who move and even by those who do not?

In this book, we find the notion of 'habitus'(Bourdieu 2002) a useful heuristic tool to examine the situatedness of migrants across scales of experience. 'Habitus' as a field of practice was developed by Bourdieu as a way of bridging across structure and agency. The habitus produces a negotiation in the field which takes shape through exchange across different types of capital – social, cultural and symbolic. It is in the field that individuals learn the 'rules of the game' in valuing different types of capital and learning how to exchange one for the other. And it is in doing so that the highly subjective values associated with different forms of capital, become objective criteria for differentiation. The field then works beyond economic exchanges through the use of networks and connections as social capital, and institutional or symbolic assets as cultural capital. This means that in different fields of practice each form of capital acquires objective value through a set of subjective negotiations, and their exchange into another form of capital is then made possible using these objective markers.

There are many critiques of habitus as a field of practice (see King 2000 for example), which suggest that in trying to bridge across structure and agency and across objectivism and subjectivism, Bourdieu actually reifies their binary oppositions. Our purpose nevertheless is to think through some of the more spatial applications of this notion of habitus. We suggest that the habitus has tremendous potential in examining the translocal 'field' of practice where migrants' (social or cultural) capital are exchanged differently in different spaces, places and scales.

While Bourdieu's notion of capital is strictly related to the social and cultural aspects of the field, Soja (2000) extends this to a 'spatial capital', which works simultaneously with other forms of capital in the field to regulate and represent new forms of practices. By incorporating space as one of the different forms of capital valued and exchanged in the field, we can think of the habitus as a spatially contingent field of meaning, working through a range of spatial boundaries, making it part of both subjectivities and physical locations (Kelly and Lusis 2006). Individuals then must learn to play the 'rules of the game' within these spatial contexts, valuing each form of capital as they become significant within different scales. These subjective locations become critical to the shaping of everyday spatial practices and hence engagements with other spatial formations (such as the home, family, urban or nation) at other scales.

Further, following on from the theory of practice in the field – learning these rules within particular spatial contexts will require successful bodily conduct, which means that habitus can be extended to provide a dynamic theory of spatial embodiment and subjectivity, through which a more reflexive transformation of identity in particular places can be realized. Individuals and groups tend to internalize the values associated with each form of capital in different spatial contexts, which means that one form of capital in a particular context will be seen as an objective marker of success in that context. And this is particularly relevant in considering how translocal geographies are also multi-scalar – in that different scales of the body, home, neighbourhood, urban, regional, national or transnational require different rules of practice which migrants must learn and internalize in order to be successful. And when translated across different scales, different forms of capital are also valued differently across different scales, which means that 'success' across one scale of the home or city, might actually be marginalizing across another scale of the national or regional.

This can be seen in a number of chapters in this collection. In Brickell's and Tan and Yeoh's chapters on the construction of a translocal home, local-local connections are negotiated across four different scales within the nation – home, neighbourhood, city and regional. In both these chapters, the migrants move between rural-urban regions, yet their construction of the local and their access to forms of capital within different scales are related to both development geographies and transnational spaces of tourism. In Hatfield's chapter on British expatriates moving to the UK from Singapore, the different scales of the family house and material belongings in this home are connected to their wider transnational move. Similarly in Datta's chapter, the intra-urban residential moves of the Polish migrants in London are shown to provide ways of accessing particular forms of capital in these shared homes that are then exchanged for economic capital at wider scales of the city. Centner notes how emplaced mobile resources connecting particular hotels and nightclubs in Buenos Aires after its financial crisis allowed particular 'immobile' social actors to mobilize forms of capital even during difficult times. And Page further emphasizes the scale of the corporeal body in negotiating forms of transport in traversing small and large distances during transnational movements of migrants between London and Douala in Dar es Salaam. Likewise, Chacko, Christou and Wise all note how local-local connections made across Washington DC and Addis Ababa; between Berlin, Athens and New York; and between Sydney and Shanghai neighbourhoods respectively provide ways to negotiate forms of social and cultural capital that can be both empowering and marginalizing in these cities. In all the chapters then, scale remains critical to construction of the local – since it is in different scales that migrants find themselves playing different 'rules of the field'.

Rethinking the translocal through a notion of habitus thus makes it possible to reconceptualize the former as material, spatial and embodied – where forms and degrees of engagement with other spaces and scales are shaped by localized contexts and everyday practices. In this way, translocality 'assumes a multitude of possible boundaries which might be transgressed, including but not limiting itself to political ones, thus recognizing the inability even of modern states to assume, regulate and control movement, and accounting for the agency of a multitude of different actors' (Freitag and von Oppen 2009: 12).

Translocal Affiliations: Spaces, Places and Connections

Divided into three main themes, the chapters in this book all consider the connections between the everyday spaces and places of the home-(land), street/ neighbourhood and city, each of which offer possibilities of engagement with mobility. Drawn to the idea of migrants negotiating what Appadurai (1996b: 47) terms, 'disjunct registers of affiliation', in this final section we outline these modes of affiliation that come together to produce multi-sited and multi-scalar translocal geographies. At the same time our objective is to show how these 'registers' are experienced in diverse forms through material as well as corporeal and subjective geographies of movement. If migration is experienced and negotiated through a variety of scales beyond the nation, then the different chapters in this book articulate precisely these dynamics between divergent scales of experiences.

Translocal Homes

The first register of affiliation this book delves deeper into is the home-(land) and its connections to different geographic scales. Home as a concept is primarily understood both as a physical location of dwelling as well as a space of belonging and identity. To date, research on migrant homes have privileged migration to the USA and post-colonial migration to former colonizing nations, tending to focus primarily on a 'loss of home' because of the nature of their migration as refugees or asylum seekers, whose return to a previous home was rendered difficult (Koser 2007). Yet, non-refugee transnational mobility suggests a more complex picture – indeed migrants are far more engaged in their own mobility, making strategic decisions in order to capitalize on their social and cultural networks and gain access to new and diverse spaces of social and political power. In this sense, we cannot simply talk about transnational movements, but need to recognize movement that is facilitated by regional citizenships of the European Union, or of migrant movements within national boundaries or urban spaces. Such a range of mobilities reflect the varieties of experiences and perceptions from wide-ranging positionalities, and demonstrate how they aid the construction and contestation of 'authentic' identities, and the formation of a number of hybridized and creolized identities around race, ethnicity, religion, and nationality, which are continuously negotiated in different spaces, places, and locations.

While the nation remains critical to much of this work, migrants' domestic homes are also connected to past, present, imagined or future 'homelands' (Skrbis 1999, Burrell 2008, Datta 2008) – perspectives which locate the home in 'a relationally linked range of localities' (Jacobs 2004: 167). The migrant home can thus also be described as translocal in ways that it is shaped by consumption, remittances, and social networks; by actual home-building, and by a range of connections to other homes in other localities. These homes are the sites where cultural difference and 'otherness' is constructed, lived, and negotiated through the ambiguous relationships between mobility and migration. In turn, these issues confront us with a number of questions which we ask in this volume – where do home-spaces end, how far do they extend, and how are the spaces between home, locale, and homeland negotiated?

In the chapters by Brickell on rural-urban migration in Cambodia and Tan and Yeoh concerning migration away from Northern Thailand, it is shown how particular individuals deal with family members separated by distance *within* the nation state. In Tan and Yeoh's chapter, for example, the use of strategies to selectively mediate distance and foster co-presence among family members are shown to be equally significant to those 'left-behind', thus evoking the idea of translocality being experienced also 'in place' as well as 'on the move'. Rather than conforming to much international development research and migration studies which tend to, according to Kothari (2008), simplify the coping strategies of the poor as overly economistic, our two chapters articulate the repertoire and mobilization of skills and expertise that allow migrants from the Global South to forge such noneconomic, social, and cultural allegiances. As Brickell shows in connection to post-migration possessions in Cambodia, evidence of domestic mobility is held in high esteem, not just in financial terms, but also as material and symbolic capital to communicate cultural engagement beyond Cambodia's borders. This resonates with what Werbner (1999: 20) also found in the context of South Asia, that 'working-class cosmopolitans', were adept in brokering knowledge of, and familiarity with, other cultures through 'specific kinds of focussed networks'. Hence rather than perpetuating a Northern-oriented slant to the study of mobilities (Benwell 2009, Gough 2008), this collection brings into view a large range of migrant actors who are actively involved in the creation of local-local connections.

This range of migrant actors also includes elites. In our volume we challenge the idea of elites' instantaneous and painless traversing of transnational space and demonstrate the situatedness of migrants for whom the contemporary world is supposed to have become 'ageographical' (Hannerz 1996; Sorkin 1992). The place-making activities of transnational (Sydney) elite (in Asia) have, for example, been shown to work against this 'thesis' of 'denationalized' identities with relocation ceremonies and the creation of expatriate spaces important to city life (Butcher 2009). Of particular significance in our collection is Hatfield's research on elite UK transnational workers in Singapore, which shows the multi-scalar nature of local-local connections. Here, transnational moves between Singapore and UK are described through local subjectivities of home symbolized through aesthetic and material evocations of views from their UK houses or their gardens next door.

In Hatfield's as well as Brickell's analysis then, the material geographies of movement (explored via participant-directed photography) speak to the complex ways in which articulations of locality and translocality can become contested terrains of home and belonging. Migrant homes and the possessions it contains are shown as highly significant expressions of power, identity and even modernity, so much so that the architecture of homes can produce what Mitchell (2004) identifies (in the context of 'monster houses' built by Chinese immigrants in an elite Vancouver suburb) as competing narratives of citizenship, democracy, and belonging. All these complex engagements between the material, symbolic, and political aspects of home ultimately then point up the ways in which articulations of locality and translocality can become contested through the register of domestic affiliation.

Translocal Neighbourhoods

The second 'register' we concentrate on is that of the neighbourhood. For Appadurai (1990) the anthropological terrain of the 'neighbourhood' – in which sedentary and circulating populations co-exist, is held up as emblematic of translocality. The term 'neighbourhood' is used to make reference to situated communities which are not only sites for the production of subjectivities but also for the initiation, enactment, performance and reproduction of meaningful social actions and activities which are informed by multiple places elsewhere (Appadurai 2005). Neighbourhoods thus function as expressions of locality through their agency, sociality and reproducibility, and which have 'both "traditional" (place-based) and "virtual" (based in communications technologies) aspects' (McKay 2006b: 199). Indeed, the significance of neighbourhoods, particular districts and suburbs has been highlighted by a host of authors as smaller-scale differentiations of transnational spaces (Ehrkamp 2006, Frieson et al. 2006). Neighbourhoods however are not just localized receptors for transnational processes; they are substantive social forms in which local subjects are produced. In Wise's chapter on the diasporic place-making activities of Chinese inhabitants living in an Australian suburban neighbourhood, this is exemplified by studying the Chinese community's cultural and material appropriate of the local streetscape. Equally, in Datta's chapter on East Europeans in London, neighbourhoods are not only shown to connect to other cities, regions and nations, but also to the everyday routinization of activities within different parts of London, activities which are compared to those of other minorities.

In this sense, we are suggesting a focus on the traditional space of the home, family, community and neighbourhood, which are often the immediate site of encounters with 'otherness' and where notions of belonging and attachment are produced. We argue that it is these neighbourhoods which could be seen as the building blocks of translocality in ways that facilitate – through kin and friendship networks embedded within them – the ability to migrate, be mobile yet maintain connections across localities. This possibility is demonstrated in Page's chapter that illuminates the dangers associated with the nation as the sole register of affiliation. In the chapter covering Cameroon inspired home-town association in London, multiple districts *within* the home and host nation are found to be of heightened significance in the individual biographies of club members. In the case of Tanzanian associations, by contrast, it is also shown how people are more likely to associate through their church or mosque than through affiliation to their hometown. Similarly studies have shown, for example how specific urban sites, like villages, also act as cultural hearts for transnational migrants to connect and organize translocality (Falzon 2003, see also work being undertaken at Queen Mary, University of London on 'Diaspora cities: imagining Calcutta in London, Toronto and Jerusalem'). We argue therefore that any focus on the neighbourhood should be accompanied by a consideration of the micro geographies (of the home and street for instance) which interact with these other locales in order for the complexities of migrant attachments and identifications between, and to, different sites to be made visible. It is only then that the richness and diversity of migrant connections with multiple places, spaces and temporalities can be more fully understood.

Translocal Cities

The third 'register' the collection engages with are cities as sites of translocality *par excellence* harbouring places of origin, settlement, resettlement and transit. Situated within the intersections between place and displacement, location and mobility, settlement and return, cities are critical to the construction of migrant landscapes and the ways in which they reflect and influence migratory movements, politics, identities, and narratives. Indeed, as Chacko's chapter emphasizes, translocality is evoked through multiple flows between Ethiopia/Addis Ababa to the United States, secondary migration between U.S. cities, as well as mobility between different neighbourhoods in Washington D.C. In turn, cities become sites of encounters with those who are different from oneself and they provide spatial contexts in which specific attitudes and behaviours towards others are produced and practised. Attitudes such as these towards 'others' are shaped by the triviality of conducting everyday practices of living and working, by 'building bridges of cooperation across difference' (Sandercock 1998). As Cohen (2004: 145 and 148) suggests in the case of non-elite cosmopolites, they 'need to know how to provide services ... need to develop foreign language skills, [and] knowledge of migration policies', which often means a 'mundane cultural interaction' with other urban actors and subsequently a heightened sense of awareness of the differences between urban localities across which they must live and work. Under such circumstances, cities, urban spaces, and neighbourhoods have become the sites of powerful and exclusionary politics around multiculturalism, inclusion, belonging, and the construction of history. In Christou's chapter on returnee Greek migrants to Athens, for example, it is shown how difficult it can be for returnees to re-adjust to national culture, on account of the alienation, exclusion and disappointment that can accompany these moves; a disjuncture emerges of the romanticized image of 'home' versus the reality of everyday life in the 'modernized' city.

As we have already intimated, we propose that migrants' everyday lives are negotiated and experienced not just at the level of the city but also within specific urban sites – in its workplaces, homes, and a range of buildings, streets and neighbourhoods where divergent and often conflicting formations of the local are produced (Datta 2009a). This is particularly evident with respect to Centner's chapter on the Argentinean economic crisis where workers in elite hotels activated forms of capital by connecting to locales abroad that are highly valued in Buenos Aires. This and other chapters in the collection thus destabilize power and social relationships from the national to a range of spaces which embody affiliation and loyalties for migrants *precisely because* of their entrenched inter-scalar relations with other more distant spaces and places. The city is a fractured collection of mundane places that produces connections (both social and material) with other spaces, places and locales within and beyond the city or nation. Moreover, just as the city shapes migrants' everyday lives, migrants too construct, rework and transform the city through their transnational and translocal mobilities across a variety of spaces and places from the church or mosque to the hotel, and from shop and restaurant fascias to residential gardens.

Towards material, embodied and multi-scalar geographies of movement

As the chapters in this book testify, translocal geographies are multi-sited and multi-scalar without subsuming these scales and sites within a hierarchy of the national or global. Exemplified by chapters which span locales within the United Kingdom, Poland, Germany, Greece, the USA, Argentina, Mexico, Cambodia, Thailand, Singapore, Tanzania, Cameroon and Ethiopia, this collection engages with the concept of translocality from the point of view of a range of different actors engaging in different geographical orbits, under different structural conditions, but all forming various social, cultural, political, corporal and agentic connections during mobility and movement.

By focussing on these connections, the collection contributes towards what Blunt (2007) and Gorman-Murray (2009) see as embodied, material and politicized mobilities which go beyond theoretical abstractions or simplified push–pull explanations. We also agree with Gorman-Murray (2009: 444) that migrants are embodied actors, and that 'sensual corporeality, intimate relationality and other facets of emotional embodiment also suffuse relocation processes'. While spaces of transcience have been linked to an absence of meaning (as argued by Augé 1995 in relation to airports), we argue that these, and other sites of movement, channel wider formations of state power, as well as more embodied, affective and emotional landscapes for migrants. Burrell (2008: 369-370) illustrates in the context of 'ordinary' migration from post-socialist Poland how the materiality of journeys and borders also 'transforms abstract transit spaces into tangible and identifiable transit places – places which are in themselves "real" points, experienced as intensely as other points in the migration journey'. In this way borders are not just physical lines in the sand but are experiential places where difference is rendered salient.

Equally, in this collection we also highlight the simultaneity of embodied and material aspects of movement that take shape in and across physical places and localities as migrants deal with issues of loyalty, commitment and emotion towards family, friends and community (see Conradson and McKay 2007). In Wise's chapter, for example, aesthetic and sensuous loyalties to Shanghai are materially evident in the street signage and restaurant window displays, all of which have affective impact, creating a sense of 'homeliness'. Equally, for Polish men in Datta's chapter, experimentation with 'otherness' in the form of Chinese and Indian takeaways is balanced against the familiarity of Polish cooking smells conjured up in their own homes. Thus, much like in Law's (2001) Hong Kong research which identified the active creation of 'home' places by Filipino domestic workers through taste, aroma, sights and sounds, this collection also touches upon what we see as the mobilization of translocal senses.

Marking out everyday vestiges of translocality, such examples hint at the 'personalised subjective temporalities' (Urry 2007: 122) that co-exist through the mobilities of daily life, as well as through larger-scale migrations. We argue that daily practices of translocality reside not only in physical movement but also extend to often mutually constitutive acts of visualizing and imagining connections between places and spaces. First, in regard to the visual, Tolia-Kelly (2004a: 681) has shown in relation to South Asians in Britain, some of whom have no chance of returning home, that a photograph becomes 'a store of social history rather than a place which is vital and present which with can be re-connected with'. While for the majority of migrants in our collection the possibility of non-return is not such a heightened issue, the visual narratives we include often relate to projected future lives 'back home' as in the case of Datta's Polish migrants and Hatfield's sample of British elite (once) living in Singapore.

The importance of the visual we argue is also inherently connected to the mobilization of the translocal imagination on a quotidian basis (see Hannam et al.

2006; Jones at al. forthcoming). In concordance with van Blerk and Ansell (2006, in relation to children), we argue that these embodied, corporeal and affective experiences in different locales have nevertheless tended to receive less attention than peoples' physical experiences of spaces. Datta (2008, 2009b) illustrates in this regard how the embodied process of house-building by Polish migrants in London actually produces particular notions of ethno-national or gendered identity that are used to imagine and materially build an 'authentic' Polish home and family in Poland (see also Datta and Brickell 2009). Likewise, Page's discussion (in this volume) of the urban landscape of Bali Nyonga, Cameroon points to the multiplying tapestry of migrant houses which have been designed and built through their owners' imaginative and literal journeys (with the attendant emotions this stirs), as well as the necessities of transnational communication with suppliers and builders.

These literal as well as imaginative links between 'heres' and 'theres' thus point to what Gielis (2009: 275) contends as highlighting an individual's 'ability not just to experience the social relations that are located in the place in which he or she is corporeally standing, but also...to experience social relations that are located in places elsewhere'. In this way, 'more important than the traversing of geography in these stories is, rather, the movement of the mind or imagination, enabling a shift of cultural and perceptual frameworks' (Wilding 2007: 332). A good example here is Wolseth's (2008: 311) research with men in Honduras who struggling with violence and diminished employment opportunities, view migration as the only viable option to escape both physical and social death. In this context, the preeminence of 'migration talk', of escaping to the United States, allows for discursive evidence of agency in the face of what is perceived of as an otherwise bleak future characterized by physical rootedness. In situations of physical rootedness, translocality then can enter into peoples' lives through a prism of possible lives offered through the imagination with fantasy used as a social practice to connect different places (Appadurai 1990). As shown by Brickell in relation to mobile phones and by Tan and Yeoh in regard to messaging services (SMS), phone calls and the Internet, the communication media also plays a distinct gate-keeping role in the construction of the translocal family and the keeping alive of domestic imaginaries and intimacies across virtual as well as geographic space.

In summary then, the approach we take to studying these different repertoires and emotions of movement is to bring into play these different sites and scales. We are not just interested in the migratory journeys that people make, but also the spatialities and pluri-localities of their lives which they negotiate on different temporal bases and through different material and affective processes. The chapters in this collection are thus both theoretical and methodological – they bring together interconnected sites that recognize the diversity of migrants' real and imagined journeys across spaces and places. They achieve three things: first they explore connections between different spaces and places which might be separated by huge physical distances; second, they examine local-local connections through a range of different scales, from the body to the nation and beyond; and finally they do this because they pay attention to migrants' embodied and material practices and experiences in a range of sites that exist within the interstices of departures and destinations. The chapters that follow this introduction thereby destabilize a cultural politics of migrancy, which solely relies on national identities and national spaces, and argues that heightened mobility and virtuality do not in any way reduce the importance of locales. And it is indeed the scale of the locale itself that is negotiated through multiple articulations of situatedness which constitute the translocal geographies of migrant lives.

PART 2
Translocal Spaces:
Home and Family

Chapter 2

Translocal Geographies of 'Home' in Siem Reap, Cambodia

Katherine Brickell

Introduction

> Locality is an inherently fragile social achievement...locality must be maintained
> carefully against various kinds of odds (Appadurai 2005: 179)

In April 1975 approximately three million people were emptied from towns and cities throughout Cambodia to destroy what the Khmer Rouge regime cast as 'parasitic locales' (and populaces) embodying a disgraced culture of modern architecture, commodities, western fashion and literature. The leader of the Khmer Rouge regime, Pol Pot, and his cadres deemed foreignness of any kind as a threat and aiming to create a 'pure' Kampuchean revolutionary worker-peasant marked a large proportion of the elite, skilled labourers and common urbanites for elimination (Jackson 1989). Using forced migration as a strategic means to disorientate and dislocate, the transportation of these populations from one end of the country to another was to become a 'hallmark' of the revolution (Becker 1998: 232). With this, homes, and even entire villages, were physically demolished to make way for communal buildings and working teams replaced families as the new basic socio-economic unit of society (Ebihara 1993). The need to destroy such personalized attachments to familial spaces of belonging was ultimately tied to a need identified by the Khmer Rouge to ensure the creation of new subjects devoid of any loyalties to the past. For the Khmer Rouge then, shared histories of family life could not be afforded. Cambodia needed to be 'killed', to cease to exist both literally and symbolically, for Democratic Kampuchea to be built (Tyner 2008: 119).

With the ousting of the Khmer Rouge by the Vietnamese in early 1979, however, hundreds of thousands of Cambodians traversed the country in search of family members and on their own came together and re-created cities and towns (Gottesman 2003: 188), initially elated as they realized they could 'go home' (Daran Kravanh cited in Lafreniere 2000: 154). Now, after decades of upheaval, Cambodians are negotiating domestic situations shaped through the country's move to market-driven capitalist growth and entry into the global cultural economy. As part of these contemporary shifts, Tyner (2008: 144) argues in the context of the Khmer Rouge genocide (1975-1979), that what is now needed is a 'more complete picture of how the people actively sought to maintain their attachment to the landscape of Cambodia and how they articulated a renewed sense of place and, thus of humanity'. While in my other writings, I have focused on how families reconstituted their domestic situations in defiance of the constraints imposed in the past (Brickell 2007), this chapter discusses the production of translocality in the context of current day tourist-oriented Siem Reap, home to the temple complexes of Angkor. It is in this UNESCO-dominated town that urban dwellers' domestic practices and strategies are to be understood both in the context of sustained connections to rural homes as well as to urban homes marked for Cambodia's future. Aiming, as many recommend, to view the Pol Pot period from a greater distance, distinguishing it more dispassionately as part of larger historical and political processes (Marston 2005: 501, see also Brickell 2007, Chau-Pech Ollier and Winter 2006), the chapter considers rural-urban migrants' evolving relationships to 'home' as a metaphorical, and potentially multi-sited space of personal attachment and identification. In doing so, I extend analyses of home which exist only from the perspective of the Cambodian diaspora in the United States (Ong 2003, Smith-Hefner 1999, Um 2006) in which a static image of home resides, one reduced and constructed 'out of treasured fragments of pre-war memory, preserved in nostalgia and kept frozen in time' (Um 2006: 89). Instead, I argue that in order to move beyond such fixed portrayals of home, translocality is an useful theoretical tool to elucidate the meanings of home that are not only being re-inscribed within Cambodian borders, but are being framed through urbanites' deployment of imaginative and material resources to transcend this very territorial marker.

Translocality as a Theoretical Tool

Elaborating on translocality as a theoretical tool, I identify four (though not exclusive) scales at which its value can be felt in the case of Cambodian rural-urban migrants. First and foremost, the very semantic basis of trans-*locality* emphasizes a greater openness to, and acceptance of, interactions *within* the nation. These are interactions in migration and mobility studies which are, in relative terms, under-theorized given the hegemonic pivot of enquiry (until very recently) being rooted in geographies of home reproduced and recast through migration between nations. Concurring with Wilding (2007), this chapter argues that alternative, or at least reworked approaches to transnationalism should focus on practices and understandings of mobility, which should include, but not be limited to, mobility across national borders. Indeed, while Guarnizo and Smith (1998: 26) argue that 'it is crucial to systematically study the translocal micro-reproduction of transnational ties', what is arguably lacking is an appreciation of the 'translocal micro-reproduction' of ties which are not exclusively transnational in nature but bound to locales within national boundaries. Exemplifying this point, Hardy (2004: 218) shows, for example, the limitations of the term 'transnationalism' through his work on the conflict between visions of the Vietnamese nation during the 1970s.

Using what he calls the 'exploratory vocabulary' of 'internal transnationalism', Hardy argues that this other framing captures the existence of two nationalisms within in the single country (namely a communist north and an anti-communist South with identifications to the West). Taking this critique even further, Collins (2009: 437) comments that 'despite the fact that the focus on transnationalism was supposed to "unbound" the study of migration from the nation-state, much work within the field operates within assumptions that continue to reinforce bound understandings of place and society'. Hence forth, this chapter contends that much wider fields of action structure contemporary migrations, so much so that tainted by a nation-centric legacy, scholarship on transnationality should be accompanied by a less haunted focus on translocality.

Having said this, taking an inter-disciplinary perspective does unmask work within development geography, which has for many decades highlighted the importance of multi-scalar connections forged through rural-urban migration, for example. As in Cambodia (Brickell 2007, 2008, Derks 2005) studies of the Southeast Asian region have emphasized the increasing role of internal mobility in supporting and defining livelihoods within some households and villages in Lao PDR (Rigg 2007), Thailand (Mills 1997, 2001, Rigg 2006), Vietnam (Resurreccion and Van Khanh 2007) and Indonesia (Elmhirst 2007). Associated with a progressive shift from farm to non-farm work as well as income earning from multiple sources, rural-urban linkages have been looked upon mainly from the perspective of livelihoods, the interconnectivities between local economies and the related flows of people, goods, money and information (Tacoli 2006). As Rigg (2005: 175) contends however, underneath this 'hardware' of development-related changes are equally as important 'software' transformations 'concerning the psychology of modernity, what it means to be a farmer, and the aspirational profile that informs and propels change'. To reflect the potential for such 'software' change, I consider migrants as interpretative subjects of their own mobility, rather than only products of broader political-economic conditions (see Silvey and Lawson 1999 for discussion of migrants as agents negotiating material and discursive dimensions of development). In turn, this chapter demonstrates that even if employment and economic imperatives are major factors, migrants' experiences are intimately interwoven with social, family and cultural considerations that influence the nature of their translocal practices and identifications.

Second, is the value of translocality for highlighting localities as specific situated places of connectivity that enable, rather than curtail, mobility within and beyond the nation. Even when migrants settle 'in place' this does not mean their lives are necessarily rendered immobile, rather their lives can be transformed without the need for physical mobility. This can be seen in a number of regards. First, tourism fosters global flows that link migrants in Siem Reap to the discursive domain of the global through exposure to foreign ideas, cultures and lifestyles. While tourism therefore helps 'to create, recreate and distribute images and flows around the world', this is not just limited to the souvenirs and photographs transported home by tourists as Shaw and Williams (2004: 6) argue, but informs the emergence of hybrid 'host' identities. In this sense, localities such as Siem Reap, not only facilitate the livelihood-related mobility of Cambodians, but also act as a confluence of global forces. Second, given the installation of Internet and mobile phone services to meet tourist and local demand, Gielis (2009) argues that this makes it easier for migrants to make contact and experience other places without physically changing location. Post-migration for example, new technology is a key means of keeping virtually alive distant ancestral homelands. In connection to second-generation Moroccan migrants in Holland, Lenie (2006) demonstrates how websites and message boards show that although young people do not maintain strong physical ties with the homeland of their parents, they do seek to *imagine* Morocco on the Internet by accessing music, sport, Islam and tourist information. Such situations thereby reflect the wider phenomenon identified by Appadurai (1996b) of increasing numbers of people viewing their lives through the possibilities offered by the varieties of mass media.

Another aspect of this 'emplaced mobility' then relates to the idea that while people can remain spatially local, their lives may also be shaped by various translocal cultural imaginaries, something that has received far less attention than peoples' physical experiences of spaces. As this chapter demonstrates, when Cambodians embark on rural-urban migration within national borders, the cultural meaning of this movement often extends far beyond rural or urban parameters to symbolically engage with notions of global modernity, which are largely de-centred from specific national territories. As Wilding (2007: 332) comments accordingly, 'more important than the traversing of geography in these stories is, rather, the movement of the mind or imagination, enabling a shift of cultural and perceptual frameworks'. Exemplifying this symbolic as well as literal movement, for instance, is Langevang and Gough's (2009: 741) Ghanaian research, where neither young peoples' daily mobility nor their spatial imaginations were restricted to Madina, 'real or imagined travel takes them to other parts of the city, into rural areas and across the nation's borders'. Thus far, the theoretical lens of 'translocal' is useful precisely because it emphasizes the links between specific localities at the same time as linking with the discursive domain of the global, but without completely losing sight of the national. Migrants' 'translocality' across villages and urban centres thereby exposes the necessity for understanding mobility both as the horizon rendering of scale, encompassing multiple locales or regions of identity, and the more orthodox vertical scale-hierarchy linking local, national and global processes (Oakes and Schein 2006).

Third, turning to the interconnected scale of the domestic and using translocality as a 'lobbying' tool, I suggest that there is great deal of mileage to be gained through interrogating transnational theorizations of 'home' for understanding other situations of mobility. It remains more broadly that the majority of theoretical work on home (and multiple allegiances) has been developed through insights drawn almost exclusively from transnational migrants (Ahmed et. al. 2003, Al-Ali and Koser 2002, Blunt and Dowling 2006, Rapport and Dawson 1998, Walsh 2006). Withstanding this bias however, the 'home' has been gainfully established as a dynamic process, 'involving acts of imagining, creating, unmaking, changing, losing and moving "homes"' (Ali and Koser 2002: 6). In turn, two main components of this theoretical work from transnational migration has emerged, one, the home as an actual place or nodal point of social relations and lived experience, and two, as a metaphorical or discursive space of belonging and identification (see Armbruster 2002, Blunt and Dowling 2006, Rapport and Dawson 1998). Such dual conceptualizations I argue can be productively grafted onto translocal geographies of home. Rural-urban movement, for example can also result in migrants being physically present in urban localities while at the same time being part rooted, often psychologically, to rural homes and family members left behind, so much so that 'home' and 'place' become 'ambiguous and shifting notions, where multiple identities – both "rural" and "urban" – can be simultaneously embodied' (Rigg 2003: 198 see also Rigg 2006).

While scholarship in transnationalism studies remains hugely relevant for thinking through geographies of home within the nation, I contend that is has largely ignored home places that fall outside its territorial remit, yet, which filter adjacent processes. Just as Conradson and McKay (2007: 170) demonstrate, for example, that 'transnational networks can be harnessed to transmit money, objects, expressions of love and support', these processes are not ones harnessed exclusively by the transnational migrant but also other mobile peoples who use sub-national networks to achieve similar goals. Material cultures are not only meaningful to transnational migrants (see Datta 2008, Tolia-Kelly 2004a, Walsh 2006) as a means to reconcile the existence of multiple homes located in different countries; they also foreground material and emotional connections between rural and urban locales. The participant-generated photographs in this paper thus build on a body of literature concerned with the meaning of domestic material cultures in relation to rural-urban migration (Hirai 2002, Mills 1997).

Fourth and finally are conceptual questions over 'homeland', a notion normatively mapped onto national space. In contrast to this trend, I argue that peoples' sense of belonging in Cambodia is not necessarily best understood through a notion of 'homeland' which privileges ideas of national sovereignty in transnationalism studies, but rather one that includes local politics and points of familial affiliation evoked more appropriately through the term translocality. In Cambodian custom, the house and village of familial origin represented a key node of belonging for Khmers in traditional society who according to Ponchaud (1989: 163) felt that 'no place else offered as much protection as the village where the "ancestors" were familiar'. Indeed, as this chapter goes on to show, despite the dislocation of Khmer Rouge years, such ancestral ties to rural locales remain of symbolic importance for rural-urban migrants residing in Siem Reap. It is in this context of an alienating state that attachments to landscapes in Cambodia are rooted to local rather than national referents. I thereby demonstrate how in the context of a genocidal regime that ultimately failed to replace home with some other collective or emotional anchoring, familial attachments to local landscapes are hugely significant in the contemporary period. Such local-based subjectivities have also been highlighted in Vietnam where families in Hanoi show a deep sense of attachment to their ancestral villages or rural 'home places' (Schlecker 2005: 510). Much like in Cambodia, for these urban-dwelling families, the word 'home place' denotes a locality or area that is essentially rural and where kinship relations are felt by virtue of one's ancestors having lived there (ibid). Taking each of these four central scales together then, translocality has the ability to reveal how migrant self-hood is formulated not only with reference to nation-states but critically also to localities within and between.

Researching Translocal Geographies of 'Home'

Under very different circumstances to Khmer Rouge times, Cambodia is now coming to terms with rapid socio-economic change. With this, after the gradual urban repopulation and reconstructions of the 1980s and accelerating developments nationally in the 1990s, home and family life is now being affected by the Cambodian government's pursuit of transformation with a great emphasis on foreign investment (at least up until the global economic crisis of 2008 onwards) in construction, the service sector, garment industry and tourism. On account of these priorities, geographical mobility in most places in Cambodia now includes significant and fast-rising proportions of newcomers especially in urban areas (Acharya 2003) as rural livelihoods diversify away from traditional farming. In Siem Reap this is particularly noticeable with the emergence of new prospects associated with tourism, making the town the second engine of Cambodia's economy after Phnom Penh (Brickell 2008). The outlined changes mean that for many Cambodians their 'translocal geographies' of home are influenced by the increased propensity for household dislocation of family members across rural and urban thresholds.

Based on qualitative research in Slorkram commune, Siem Reap, this chapter emerges from a larger project concerned with understanding domestic gender relations in the context of Cambodia's macro-economic transition. Within this project, I conducted a combined total of over 160 discussion groups, oral history and semi-structured interviews with respondents ranging between the ages of 20 and 70 (including rural-urban migrants in this chapter). Twenty of the individual histories included participant-directed photography. The recent migrants I draw upon were not seasonal migrants but engaged in a wide range of occupations throughout the year relating directly to tourism, including waitressing, taxi driving and souvenir selling. The chapter also includes a number of older migrants of rural origin who settled in Siem Reap in the 1980s and 1990s undertaking occupations unrelated to tourism and who, in some instances, are now retired.

The interviews that included participant-directed photography were held in two stages. In the first, I learned more about respondents' lives (and migration histories) and explained the logistics surrounding the photography exercise. This involved provision of disposable cameras to participants, gaining informed consent for using the photographs in my future work, as well as briefing them that they were to take photographs of anything they thought was important, significant or interesting in their lives. In the second sessions, it transpired that around one third of the photographs contained images of domestic possessions (each participant took approximately 10-15 photographs in total). The inclusion of home possessions is not necessarily surprising given that the home is a critical site for material culture (Miller 2001). Given my use of a Khmer research assistant who translated and transcribed the interviews into English, the photographic resources were also helpful personally to establish a shared local verbal and visual vocabulary within the interview encounter.

Mobilizing the Translocal Home in Contemporary Cambodia

The empirical element of this chapter is divided into two main sections, the first of which considers how the translocal home is being mobilized as a present-day strategy. Given the multiple forms that translocality can take, in the second section I go on to question what these geographies of home mean in relation to participants' perceived futures. In regard to the present, the findings show that while participants use translocal networks to sustain the dual liminality of their domestic existences, to obtain urban employment in the first place, migrants must tactically use multi-located patronage networks. In Cambodia, it is especially important to have *Ksae*, translated literally as 'string', 'rope' or 'line' (Derks 2005: 88) running between, and centred within, both rural and urban locales.

This can be clearly seen in the case of Channery who used one such network to secure employment at as waitress. Previously a farmer, Channery migrated from Kandal Province with a friend in 2003 and left her only son, Sok, at home with her mother. It was the first time that Channery had migrated, except at a young age during the Khmer Rouge when she was evacuated to Battambang Province for three years. Facing economic difficulties following her husband's desertion to marry another woman, Channery was drawn by the perceived opportunities tourism held for service-related work, the income from which goes towards her fourteen-year old son's education (who she felt would receive a more stable background with her mother because of her own anti-social hours of work as a waitress).

Once in Siem Reap, the mobile phone represented a key mediator in Channery's translocal interactions allowing many other migrants like her the means to maintain and explore new social networks in the town at the same time as communicating beyond it. As Gielis (2009: 280) notes more widely, the revolution in computer and communication technologies since the 1990s has meant that 'migrant places are no longer localities, but have become translocalities – places where transmigrants reach out to other places'. In Channery's instance, the mobile phone offers a powerful form of solace, functioning as Castelain-Meunier (1997) argues in the context of divorced and separated fathers from their children, as a kind of umbilical cord linking the two in a form of virtual parenting. Indeed, as Miller

(2009) prompts, the mobile phone should not be viewed as merely incidental to migrant relationships but profoundly instrumental to sustaining the possibility of a translocal home. Deemed her most cherished possession, Channery chose correspondingly to take a photograph of her mobile phone.

Figure 2.1 Channery's picture of her mobile phone

The mobile phone was particularly important to Channery as a means of overcoming her restricted mobility, as she could not afford to visit her son often for fear of reducing her salary. In this way, the mobile phone represents but one of the communication methods that migrants, and particularly mothers use 'to develop intimacy, in other words familiarity, across borders' (Parreñas 2005b: 317 in the context of Filipino transnational families). Sitting by her bedside, Channery elaborated,

> I miss my home very much…every day whether it be sleeping, walking or sitting, I always miss my old house. I miss my son and I worry because he lost his father when he was young and he hasn't the comfort from his mother either because I have left him. Sometimes I think I will go crazy thinking about him. When I go

anywhere I miss him and everyday all I do is for him. I brought only my son's photograph because it is very important to me. I keep the photo at the bedhead because if I keep it at the head you will know that I love my son. Although I am active, I always carry it on my mobile phone, call him everyday, and sacrifice all I can for him. [Channery, 28 years old, deserted by husband, waitress]

In light of the emotional pain Channery experiences living apart from her son, she attached a photograph of Sok on the back of the mobile phone and carries it with her at all times, partly as a deliberate attempt to signify her fulfilment of a key status requirement for Cambodian women – motherhood (see Brickell 2011 for further details). As Rose (2003) aptly contends, the more distant people are the more important photography is, bringing absent locales and family members into view. Channery's experiences therefore point to the significance of often-neglected emotional geographies of translocality, which preoccupy migrants and their attempts to deal with the spatial splintering of households in the contemporary period. Tensions in this regard we also found by Setha who in two separate instances explained the difficulties of sustaining his family's translocal existence,

When we talk about the house, we could be talking about the house in one's hometown, the house you rent…but you will get different feelings. Most people don't think they live in a home. My wife doesn't feel at home living here – she wants to go back to her rural homeland where her family is. So I say to her- are we going to split from each other just because you want to live in your homeland? [Setha, male, Slorkram, 48 years old, married, doctor]

As Setha's interview points out, for his wife, 'home' still means her homeland and the house she now lives in, is just a surface shift. Echoing the idea of homeland being located *within*, rather than *of* the nation, the perspective of Setha's wife show how 'homeland' can be related to specific places of belonging. Furthermore, while people re-orientate themselves not only to their new homes, but also to their old homes and memories (for many by returning to festivals such as Khmer New Year and *Pchum Ben*, the festival of the dead), it remains the case that despite rural loyalties, migrants such as Setha also form commitments and attachments in their immediate local contexts. As Setha explains,

You get used to moving around from house to house and learning new things. Your perspectives are different moving from and to varying places, at different times and under different circumstances. You learn something from different people, cultures and lifestyles. For example, in Pursat and Battambang I lived under the same regime yet had a different perspective on life management. I grew up moving – it's hard to adapt but you have to make something special at every place – in your environment. In some of the places during your movement you cry…I meet someone special there – my wife – a friend – your neighbours almost everywhere I go I turn my mind to meeting people. When you move you make something special in each place…these connections remain with you forever. [Setha, male, Slorkram, 48 years old, married, doctor]

As Setha's interview suggests, the continual movement resulting from the Khmer Rouge's strategic use of migration has created a situation where for some, movement has become an integral part of their psyche and identity. Through his comparison of different places, Setha demonstrates 'home' as one point in a more dispersed geography of dwelling. This spanning of the family unit across provincial borders does not mean its members are entirely disembedded from locality however, rather in each place it is possible to form something 'special'. Such places are settings for interaction, meetings and personal development, experiences and connections that then inform those in other localities either concurrently or in the future. Here then, translocality allows for the local grounding – or emplacement – of the growing number of mobile subjects characterizing contemporary Cambodia.

Imagining a Translocal Domestic Future

So far, this chapter has focused mainly on migrant localities as sites that radiate current and past connections. As Gielis (2009) highlights, another important way to experience other places is to dream, daydream or imagine them, much like Channery did through the viewing of photography (see also Appadurai 1996a, Morley 2000). I contend that in order to understand translocal geographies of home, projections of an imagined and desired future are playing an ever increasing part of Cambodians' attachments, and uses, of home as a metaphorical and physical domain of familial aspiration. Here peoples' everyday thoughts now involve imagining new possibilities for themselves and those of their children. As Appadurai (2005: 54) explains, here the lives of low-income families can be radically re-imagined, no longer seeing them 'as mere outcomes of the giveness of things, but as the ironic compromize between what they imagine and what social life will permit'. This is very much the case for Yan, who tired of travelling to Siem Reap to sell his sculptures to tourists moved in 2002 with his family to the town to take advantage of the rapidly expanding tourist souvenir market. The father of three explained however that he did not want to teach his children to carve (unlike his father) rather he viewed his children's future to be governed 'by paper' (education) rather than by 'wood' (carving). Aligning the accumulation of social capital with the material cultures of technology, Yan captured this in a photography taken of his eldest daughter surfing the Internet.

Figure 2.2 Yan's picture of his daughter on the Internet

Yan explained that he took the photograph of his daughter in a guesthouse she worked in part-time, employment that the family valued economically, but also enabling Melea to practice her spoken English and complete her school homework by researching online. In this way, Siem Reap represents for Yan's family what Oakes and Schein (2006) see as a place where people do not need to move in order to have their subjectivities transformed by translocal processes. Commenting on this process, Yan explained further that,

> I feel that foreigners would like to come and visit Angkor Wat and so I am
> very proud as a Cambodian. My family's life has changed living in Siem Reap
> because here it is easy to sell the sculptures and my children can learn English,
> Thai or Japanese and find a job working with foreigners to generate an income.
> [Yan, male, 60 years old, married, stone carver]

As Yan intimates, Siem Reap is not only a town where different nationalities converge but also where Cambodian's can feel proud of their own cultural heritage. In fact, Angkor especially has widespread populist appeal and iconic status as a national and ethnic symbol which attracts Cambodians, particularly at Khmer New Year, when the monument comes to represent a recovery from the social, political and economic oppression, which characterized the country's recent past (Winter 2004). At the same time as heralding a glorious past, Angkor has largely facilitated a global future for the town (albeit with the economic downturn hitting visitor numbers). With this, geographies of home are characterized by the enactment of deeply felt desires to participate in what is perceived as Cambodia's global development, with Angkor spearheading this. In Siem Reap, many migrants therefore see themselves as 'mobile subjects who draw on diverse assemblages and meanings to locate themselves in different geographies simultaneously' (Inda and Rosaldo 2002: 19). Piseth captured a dual yearning and cross-referencing of his rural past and urban future in an ox-cart wheel garden feature he constructed in his front garden.

Figure 2.3 Piseth's picture of his garden feature

After being displaced from Siem Reap at the hands of the Khmer Rouge, and forced to work on its rice fields, Piseth worked with the Vietnamese as a soldier until 1991. He then re-settled in Slorkram as a motorbike driver and built his current house in 2000. Piseth explains his resultant construction project as follows,

> This is the ox cart's wheel that I took from the countryside because I saw them in
> hotels and restaurants and I like it. It reminds me of the countryside when I used
> to sit on the buffalo cart or oxcart by the Tonle Sap lake. [Piseth, male, 44 years
> old, married, motorbike driver]

Piseth went on to explain his inspiration for the sculpture, referencing similar decorations in hotel and restaurant complexes which he saw as imaginative reworkings of old agricultural machinery only possible in urban settings which accommodated a hybridity and modernity of thinking. In this way, while commercialized tourism 'promotes consumption of fleeting images, experiences, and sensations patched together in the collage-like, pastische effects' (Kearney 1995: 555), such processes can also influence tourist hosts who incorporate and blend their own experiences of different places. The upturned wheel not only symbolizes Piseth's current life in urban tourist-oriented Siem Reap, but also represents a vessel for memories of his past life in the countryside, leaving a 'trail of collective memory about another place and time' but, also, creating, 'new maps of desire and of attachment' (Appadurai and Breckenridge 1989: i). These new maps of desire are reflected in the English naming of his 'Lucky House'. As well as organizing the provision of motorbike taxis from Siem Reap International Airport, Piseth and his family were enthusiastic learners of English and rented the first floor of their house to two successive British expatriates working in Siem Reap who enjoyed reading foreign newspapers and television programmes (pursuits he would often share with them).

These transnational connections were also displayed through other photographs Piseth took of the first floor balcony of his house and one of his tenant's commissioned furniture, which he considered quite atypical for Khmer style homes. While Piseth is actively involved in brokering relationships across rural and urban locales, he is also simultaneously aware of the power that lies in his ability to profit financially and culturally from his connections to the 'outside' world. In this way, the urban locale of Siem Reap is again perceived positively as a key pole of tourism facilitating Cambodia's development as a market economy, and thus driving the potential for inhabitants to obtain the desired material and social capital to rival other more developed countries in the region. In fact, many men unlike Piseth felt they must distance themselves from their agricultural roots, traditionally associated with an image of peasants as poor and backward people (Vickery 1984) to do so. This situation is echoed in wider literatures on Southeast Asia as farming has become a low status occupation and in many cases 'the ideology of modernity has reworked the idea of agriculture and the value attached to farming in the eyes and minds of many' (Rigg 2005: 176). As seventy-two year old Kiri explained aspirations for social mobility are often expressed in terms of consumerist culture available in Siem Reap,

> Men want to be happy, drink, dance and be fashionable. We never thought about
> these things before now- before when we were just farmers. It is changing as

more and more people come to Siem Reap and bring new ideas. Neighbouring
countries and the rest of the world are modern, have new fashions and materialism
– we wish for that too. [Kiri, male, Slorkram, 72 years old, fruit tree grower,
married]

As Kiri intimates, tourism in Siem Reap has become an important factor
in men's longing to follow what are perceived as modern lifestyles based on
adherence to fashion and materialism and thereby develop like neighbouring
countries (most notably Thailand). The (dis)connections narrated between rural
and urban are therefore not just the product of changes internal to Cambodia
but relate to processes that many male participants are particularly attentive to
beyond Cambodia's borders. In this sense, the findings are consistent with that of
Salih (2001) in the case of Moroccan migrant women whereby a similar twofold
practice emerges: on the one hand, women appropriating and negotiating symbols
of modernity by interpreting and attributing value to goods that flow from Italy
to Morocco (and vice versa); whilst on the other hand, using these same goods
as a form of distinction and affirmation of difference with respect to those who
remained. The significance of material culture as a mechanism for conveying
social and cultural capital of a global nature is also captured in a photograph taken
by Dith (Figure 2.4). For wealthy urbanite Dith, photographs convey a sense of
'presence in absence' given that he has two daughters studying in Japan and France
as well as family members who remain in France having fled immediately before
the Khmer Rouge assault on Phnom Penh. Their mobility, which has resulted in
his own overseas travel to visit them, is re-lived and evoked in a wall-mounted
picture of his family in front of Notre Dame Cathedral in Paris.

The photos are souvenirs from France on holiday. In that one I am with my wife
and we both have lovely smiles showing how much we love and care for each
other. The pictures are important to show my children so that they can see their
relatives even though they aren't here. These pictures also show that I have
visited France. It is to tell me children that they too should go there because of its
beauty and highly developed nature. If they want to see the real view they must
save money by working hard and studying more. Travelling is good because we
can share ideas and see the new culture, watch how they act and what they do.
[Dith, male, Slorkram, 66 years old, married, retired actor]

As 'every photograph is a certificate of presence' (Barthes 2000: 87), cultural
capital is acquired not only by displaying love between husband and wife but
also through showing connectedness to places and family beyond Cambodia.
Lozada (2006: 95) notes in the context of China for example, that 'taking a family
photograph with dispersed members provides a physical manifestation, as an
object of material culture, and as a performative reminder – through its display
and viewing by family members and friends – that the deterritorializsed family can
remain connected virtually'. Although the photograph from France does not show

Figure 2.4 Dith's picture of Notre Dame

these extended family members, an important process in the interview situation was to enquire further that the person who took it was indeed his Paris-based brother. In this manner '"homes" always involve encounters between those who stay, those who arrive and those who leave' (Ahmed 1999: 340).

Stretching family integration beyond the immediate household, Dith's photograph – of a photograph – thus bears witness to Dith's connectedness. In addition to this, Dith also took a photograph of an Eiffel Tower plastic souvenir with its display mirroring Urry's (1990) contention that acting like a tourist 'is one of the defining characteristics of being "modern"'. Photographs and souvenirs from foreign places were typically imbued with positive associations tied to open-mindedness, education and self-progression. As a result, despite the obsession of the Khmer Rouge to erase a modernity fashioned on the West, the home in Dith's case is an expression of connectedness, and modernizing discourse in which evidence of global connectedness is held to signify modern success and social status. Such findings echo with those of Hughes (2004) in the context of returnee expatriate Khmers in the 1990s who identified their own personal success not with flexibility, but through their combining supposed 'Western success' with notions of essential 'Khmerness'.

Concluding Thoughts

The home is a space through which many trajectories pass. Translocal in nature, many Cambodians identify with more than one locality. In this regard, for (newly) settled urbanites, attachments to rural Cambodia are not only premised on maintaining contact with current family members left behind but are also saturated with emotion on account of past ancestral heritage and ties. Such ties between localities are being negotiated and recast as Siem Reap itself is transformed through international and domestic tourism, bringing with it new lifestyles and desires which draw their inspiration from sources originating beyond Cambodia's national borders. As part of this simultaneity, the chapter captured how images and portrayals of 'rural' and 'urban' are mediated through the materiality and symbolic qualities of home possessions, which transcend purely 'rural-urban linkages', but extend again to the traversing of the mind and imagination across times and spatial scales within and beyond Cambodia. In such a transitional socio-economic situation, power essentially lies in the ability to broker relationships across the boundaries between the inside of a locality and the 'outside world' (Castells 2000). Indeed, despite Khmer Rouge attempts to ex-communicate the wider world, this chapter has shown the significance of symbolic and material indices of translocality which are brought to this task and to which propel mainly male aspirations for the future. At the same time however, the chapter has shown the very real and pragmatic tensions that exist, mainly for female migrants, as they organize family life around their multi-sited existences. In this sense, just as temporal as well as scalar considerations are critical to understanding domestic geographies of mobility, the differential management and meaning of the translocal home is an issue that is deserved of greater scholarly attention.

Acknowledgements

The author wishes to thank the Economic and Social Research Council (ESRC) for the doctoral funding on which this chapter is based (PTA 030-2002-00869) and the British Academy for the time to write it (PDF/2008/86). I am also indebted to the interviewees and research assistants in Siem Reap who generously gave their time to facilitate the research and Katie Willis for comments on the initial draft of this chapter.

Chapter 3

Translocal Family Relations amongst the Lahu in Northern Thailand

Brian A.L. Tan and Brenda S.A. Yeoh

November. The almost month-long vacation in Thailand had flown by and it is time for reluctant goodbyes to be said once again. Ca Nart starts the engine of his motorcycle, checks his wrist watch, and calls out to his daughter who is busy ensuring that she had not left anything behind. Ca Nart's wife replies for her daughter, 'Soon soon, she's coming soon'. Ca Nart is anxious about his daughter arriving late in school as the journey to the student hostel in town will take about slightly less than three hours while he has to be back in the village for a meeting with the village head in the afternoon. Ca Nart's daughter emerges from the house with two bags, one for her books and the other rather big bag with her personal belongings in it. She hops onto the motorcycle and waves goodbye to her mother who is standing at the entrance of the house. Ca Nart reminds his daughter to hold on tightly as he pushes off and rides onto the main road leading out of the village. It would be another few months before Ca Nart's daughter returns to the village.

The above snapshot is becoming part of everyday life amongst the Lahu people in Northern Thailand. The rural village locality can no longer be taken or understood as a given territorially bounded social reality. Ca Nart's once isolated rural village is transforming into one which is increasingly connected into the urban centres of the region. With this connection, the socio-economic and political milieu of the rural Lahu village are increasingly being thrown into flux and new understandings of what the 'local' means and how it is reproduced are constantly being negotiated.

Translocal movement amongst the highland people group called the Lahu in Northern Thailand has been a way of life for many generations. In recent years, with the spread of developmental projects spearheaded by the government in the Lahu highland areas making available electricity and a wider range of communications technologies, movement from the rural areas to the towns and cities has increased in significant ways. In this context, everyday life in the Lahu village has been transformed, becoming a product of negotiations between translocal influences and local forces brought into interaction through multiple linkages at global and national scales.

Specifically, this chapter focuses attention on the scale of the 'family' as a site of negotiation of these different spaces and scales. It argues that the 'left-behind' family in the villages which have experienced outward movement of younger people to the towns and cities as places of greater opportunities for work or study is not necessarily incapacitated and should not be understood simply through idioms of loss, desertion and stasis as often found in popular accounts of these 'left-behind' communities as left without younger blood to tend the fields. Instead, 'family' is actively produced in the translocal village through everyday spatial practices and relations, drawing on both imagined and material ties that bind, and using strategies that selectively mediate distance and foster co-presence among family members. The translocal family hence draws strength from a web of translocalized relationships and transactions between the village, the city and the nation, allowing recuperation of notions of locality, not as an about-to-be-erased transient place or a residue place left over from earlier times, but as a significant site of negotiated understandings of home, family and locatedness. Translocality in the 'left-behind' village (Figure 3.1) reflects the materiality of family relations and the situatedness of the family in place, as well as its connections to the cities

Figure 3.1 Approximate Location of Research Areas

and towns where its younger people work and study. This troubles commonplace understandings of the left-behind village as disembedded and forgotten, and instead advances notions of the village as active, viable and an important node in the web of translocal connections.

The 'Translocal' Village

In what Trouillot (2003) calls a 'fragmented globality', mobility has provided new ways to envision distance through time, space, society and culture. 'Place is usually considered the backdrop for motion' (McKay 2006b: 197) as transnational migration and other global forces produce porous places which are linked to a larger network of relationships. Doreen Massey (1993, 1994) posited that localities are always temporary and contested, and emerge through co-presence or from face-to-face encounters with others at specific points in time. Therefore place, for Massey, is a product of social relations and expressions of identity, as well as a node in a larger network of other places. The 'local' is linked to other locales through social relations, an important aspect in the formation of translocalities as elaborated by other scholars.

Appadurai (1991, 1996a), for example, argues that global or cosmopolitan processes now flow through the local, that is, through 'the most localized of events' (Fox 1991: 12). He sees locality (Appadurai 1995) as made up of relationships not necessarily bounded by space or place. These localities are places where there are a whole myriad of connections to the global through marriage ties, work, businesses, leisure, the media and the circulation of people, challenging traditional territorial definitions of the nation-state. These translocal connections across space may also be 'based on imagined identities' (Low and Lawrence-Zuniga 2004: 299), producing 'populations with various kinds of "locals"...[and] creat[ing] localities that belong in one sense to particular nation-states but are, from another point of view, what we might call *translocalities*' (Appadurai 1995: 216, and 2004: 339; emphasis in original). As such, translocalities are places which have both a sedentary and circulating population (Appadurai 1991, 1995, 1996a; McKay 2006a) which is linked to the global world mediated by, for example, mobile phones (McKay 2006b). Following Brickell and Datta's (this volume) critique of the tendency in current work on migrant mobilities to privilege the national scale at the expense of other scales and spaces, we also wish to use notions of translocality to refer to spaces and places which are connected and produced by a range of dynamics *below* (but also connected to) the national.

In this context, we build on Velayutham and Wise's (2005: 27) notion of the 'translocal village' which in their view is 'a distinct theoretical subset of the broad category transnationalism', but based on the micro-scale of the village and place-oriented familial (and community) relationships which have expanded in subsequent years or decades across space. We depart from Velayutham and Wise's framing of the 'translocal village' as a distinct theoretical subset of transnationalism but share their emphasis on the need to go beyond conceptualizing the village as an isolated place or identity void of exterior influences or connections. Instead, the notion of translocality as applied to the village, places emphasis on the everyday relations which allows us to 'envisage the everydayness of material, family, social, and symbolic networks and exchanges that connect [places]' (Velayutham and Wise 2005: 40; see also Vertovec 1999) in a non-hierarchical manner and across multiple spatial horizons. Focusing on the emotional and material character of relationships within the messy dynamics of translocal migration and movement allows us to recover significant insight into the production of specific locales – in this case, the village – as embedded in complex and interconnected social worlds.

In their paper, Velayutham and Wise (2005) use the concept of the translocal village to describe a set of village-like social relations and activities which have been formed among transnational migrants in the receiving country and which mirror the relations of the 'home' village. Focusing on the translocal effects of affective moral economies underpinned by pride, shame, guilt, responsibilities and obligations to family and the village community back 'home', they showed how migrants in the destination country continued to be caught up in a ramifying network of translocal social relationships which orientated them towards the sending community and kept them within the compass of the 'home' village's social and moral gaze. The conceptual framing of this chapter draws in part from Velayutham and Wise's (2005) notion of the translocal village but focuses instead on the 'left-behind' village in the context of rural-urban (as opposed to transnational) migration. We treat the Lahu 'left-behind village' as a translocality comprized of the varied social and economic networks formed as increasing numbers of the children of the parent generation migrate out of the village. We demonstrate that the sustenance of everyday socio-economic relations as played out at different sites of significance to the left-behind family and community is predicated on the variegated web of networks, relationships and interactions that reach beyond the village. In contrast to Velayutham and Wise's emphasis on (transnational) migrants as well as most literature based on transnationalism/translocality frameworks, we take a different direction by focusing on the lives of those who do not move but who are in no way 'immobile'; instead, their lives are intimately connected to and interdependently woven with the lives of those who do. In short, we are interested in exploring the negotiation of translocal subjectivities (within the framework of the family) as they are applied to left-behind non-migrant family members, who are equally affected by geographical mobility as those who move.

'Left-Behind Families' in the Translocal Village

A key body of literature focusing on the links between 'family' and 'migration' has in recent years developed around the notion of the *transnational family*. This concept is integrally tied to transnational migration, pointing to the growing phenomenon of family members being dispersed across national borders in different countries while continuing to share strong bonds of collective welfare and unity. In the context of 'Asian' transnational families, Yeoh et al. (2005) discuss three significant themes that characterize these families: first, apart from material bonds, transnational families share an imaginary 'of "belonging" which transcend[s] particular periods and places to encompass past trajectories and future continuities' (Yeoh et al. 2005: 308); second, the fostering of intimacy and engagement among family members across space through various mediums of communications including messaging service (SMS), phone calls and the internet; and third, the strategic intent to improve their lives and carve out a better future, whether by allowing one or more family members to take advantage of new economic opportunities abroad, or as part of a 'class' move to acquire social and symbolic capital and value elsewhere.

While the three themes of imaginaries, intimacies and strategic intents forming the connective tissues between dispersed family members have emerged primarily with research on *transnational* families, we argue that they are also relevant to the case of families where members are separated by distances not across political borders but within the nation-state. Insights from research on transnational families may be adapted to apply to 'translocal families' who also negotiate distance and boundary lines (e.g. the rural-urban divide) in maintaining and imagining 'family'. For the purposes of the arguments we wish to make in this chapter, we have chosen to focus on what is usually referred to in the literature as the 'left-behind family', a specific form which may be subsumed under notions of the 'translocal family', but which gives more weight to the perspectives and dynamics of significance in source communities. This is a deliberate strategy on our part as we wish to redirect the analytical lens to the way articulations of situatedness are negotiated in source communities, given the overwhelming imbalance in the current literature in favour of migrants and their encounters with place at destination.

The term 'left behind' is a contentious one as it is structurally tied to migrants (as the active term) and associated with negative connotations, implying non-mobility and suggesting a form of dualism dividing migrants from non-migrants. In recent years, the concept of the 'left behind' has been given greater critical attention and its usefulness in focusing attention on source communities in migration processes acknowledged. Some scholars (Yeoh et al. 2002; Toyota et al. 2007) have been calling for new research frameworks that deal with migration and the left behind in a holistic fashion in understanding social change, through what has been termed the 'migration-left behind nexus'. This vein of work has moved beyond economic aspects of the relationship between migrants and the left behind (e.g. the considerable work on remittance flows) to encompass social aspects such as the sustenance of family ties (Asis 2002, 2003, 2006; Parrenas 2003, 2005a; Elmhirst 2007; Knodel and Saengtienchai 2007) and giving voice to the formerly agency-less left-behind individuals through methodological avenues of inquiry such as ethnographic writings (Parrenas 2003, 2005a). Rigg (2007: 173) argues that 'rather than attempting to predict or come to some generalized conclusion regarding the impact of migration on source communities and the left-behind, or indeed on migration more generally, a more productive line of enquiry would be to identify and think about 'lines of influence' and to study the interactions between the left behind and migration in the broader social, economic and political contexts. Meanwhile, Xiang (2007) reminds us that there are 'fundamental institutional constraints' governing who becomes the migrant and who is left behind.

The existing literature on left-behind families has tended to focus primarily on the welfare of vulnerable and dependent members such as young children (Asis 2006) and the elderly (Knodel and Saengtienchai 2007). In this chapter, however, we turn attention to the perspectives, negotiations and strategies of left-behind family members who have featured much less prominently in the literature – who are also socially and economically active members. Through their eyes, we examine the nature of parent-child relations in the fast evolving translocal village. After a brief section describing the context and methods used in the study, we discuss three main themes: how care and intimacy are increasingly transmitted through routinized translocal communication between parent and migrant child; how the left-behind family in the village continues to valorize aspects of close proximity and former ways of life, and in doing so, render cogency and stability to the lives of left-behind family members and the village community; and how the importance of the local is continually reproduced and reinforced through return visits, nostalgia and adopting surrogate relationships.

Context and Methods

Northern Thailand is inhabited by a myriad of highland indigenous ethnic groups commonly referred to by the Thais as *chao khao* or highland people. One of the many ethnic groupings within the area, the Lahu, is classified as a migratory tribe (Aguettant 1996) as they have been migrating southwards for many generations for reasons such as conflict and the search for arable land (Walker 1995, 2003). While swidden agriculture (a form of shifting cultivation) has characterized the livelihood of the Lahu people for centuries, this system has declined in the last three decades due to many reasons (Aguettant 1996) such as the expansion of logging activities and growing populations; and prejudices against what is considered an unsustainable form of agriculture (Dze 2005; Laungaramsri 2001, 2005; Sturgeon 2005). Hence, in recent times, these 'constraints' and economic pressures have driven the Lahu people to change to a more sedentary form of agriculture (Keyes 1987; Walker 1995; Nishimoto 2003; Dze 2005; Laungaramsri 2005). While some continue with this mobile form of agriculture within a reduced land area and fallow period (Walker 1975), others have moved to the urban areas in search of wage labour or joined the tourism industry, hence becoming increasingly dependent on the market economy. The latter is clearly the case in the four Lahu villages in the Chiang Mai and Chiang Rai provinces included in this study.

There is no doubt that modernization has made significant inroads in these villages. Village life is now much more inter-linked with the 'outside' world and in particular the towns and cities in surrounding areas. This is in part a result of the national population registration exercises conducted by government agencies as well as development initiatives implemented by the Thai monarch, the government and non-governmental organizations (Aguettant 1996; Baker and Phongpaichit 2005). Today, as a result of the government's school-building programme in the highlands, most Lahu children have the opportunity to attend local schools which use Thai as the language medium in their village or nearby villages (Baker and Phongpaichit 2005). Electricity from the national grid has reached some of the highland villages (this is the case for some of the villages in this study) while others rely on solar panels sponsored by the Thai Royal Project initiated by the King. The media[1] and other forms of modern influences have become pervasive, educating the Lahu people in the 'Thai way' (Baker and Phongpaichit 2005).

The population of the four Lahu villages where this research was conducted varied from around 30 to 200 households. These four villages are linked by the Chiang Mai-Chiang Rai highway, a major arterial road linking the two northern cities. In terms of livelihood, the majority of the village population engage in agricultural related activities, although there is increasing evidence of a move to depend on the market economy for survival.

The main ethnographic methods used during the course of the research were semi-structured interviews and focus group interviews. This qualitative form of enquiry is most appropriate in the village environment as the community is a close knit one where villagers' homes remain constantly open to the rest of the community. If there are any activities occurring at a villager's house, the rest of the community is free to drop by and join in. As soon as the interviewing started, some family members and neighbours of the interviewee may join in the conversation, and in some cases, this proved to be rather helpful and enhanced the whole process of enquiry by reminding the interviewee of past events or encouraging contributions of different pieces of information or viewpoints related to the topic under discussion. Some of the most important data collected were the result of informal sessions and conversations with some of the 'guests' who joined in, and who themselves became interviewees.

---

1   In the context of Indonesia, Chu and Schramm (1991:5) argued that 'the mass media as one of the major channels by which people learn about the modern world and acquire a mobile personality...which may be essential if they are to cope with new opportunities and challenges [emerging as a result of economic development]'. Their study found that television in particular is an important medium used by the government to promulgate development and unify the country. In a similar vein, in the case of Thailand, Baker and Phongpaichit (2005) wrote about the significant role of the media in broadcasting official views of national development. While we will not be focusing on the influence of the media in this chapter, there is no doubt that the mass media has an important place in the proliferation of modernity in rural communities.

A total of 40 interviews were conducted (in a mix of Thai and Lahu[2]) with left-behind parents who had either children studying in secondary and tertiary institutions located beyond the village, or adult children who were independently earning a living in the towns and cities. Most in the latter category were involved in economic activities such as teaching, selling noodles, and in construction and sales related industries in the urban centres. The left-behind parents are themselves economically active, with a large majority working in the agricultural sector (includes selling their crops in the market) and a smaller proportion working in the area of making and selling traditional handicraft products.

Negotiating Care and Intimacy through Translocal Communication

Parrenas (2005b) defined 'transnational communication' as the flow of ideas, information, goods, money and emotions between migrants and their left behind families. While her focus was on issues that span across national borders, similar issues of communication between family members across distance are relevant in the context of rural-urban migration explored in this chapter. While there is now a rich vein of scholarly work (see Hondagneu-Sotelo and Avila 1997; Levitt 2001; Asis et al. 2004; Parrenas 2005b) focusing on female labour migrants working away from home and who perform 'transnational' or 'long-distance mothering' from afar; little has been done on how left-behind parents conduct 'translocal parenting' from home sites in the village.

Parenting in the translocal village is undergoing change as 'traditional' forms of communicating with children are 'contested by alternative values that must be rationalised and defended' (Magowan 2007: 460). Acts of care and intimacy[3] traditionally performed within the confines of the home between parent and child now have to be transmitted through 'virtual contact' (Baldassar 2007) using telephone lines or SMS texts, and so on. From our interviewees, we learn that telephone calls back home was the most regular means of communication between the left-behind parents and their migrant children. Telephone lines provide a channel for building intimacy across space because language, spoken

---

2   Interviews were conducted by the lead author Brian who speaks Thai. A translator was used for Lahu interviews.

3   In this study, we have chosen not to focus on left-behind parents who are in need of physical care because of illness or old age. In these instances, scholars working in different contexts have highlighted the key importance of geographical proximity in providing care and support to ageing parents (Lin and Rogerson 1995; Bengston and Harootyan 1994) given that physical care-giving requires a measure of close proximity. This led Baldassar and Baldock (2000: 62) to conclude that older parents without adult children in close proximity, actually 'miss out on an important caring relationship and source of support'. In a similar vein, as Aldous and Klein (1991) found, it is normally the case that the adult child with the closest geographical proximity assumes full responsibility of caring for his or her parent.

or otherwise, is primarily a way we reveal ourselves to each other in deepening and continuing intimacy (Peterson 1997). As Levitt (2001: 22) argued, 'New technologies "heightened the immediacy and frequency of migrants" contact with their sending communities and allow them to be actively involved in everyday life there in fundamentally different ways than in the past.' During conversations with their children, left-behind parents generally kept their conversation topics around three main recurrent areas – first, they would enquire about their children's well-being; second, they would communicate (or reiterate) instructions about living away from home; and third, they would inevitably ask their children when they would be returning home, either permanently or for a visit, in the optimistic hope that the answer would be 'soon',

In the majority of cases, family strategies of care and intimacy have readapted and both sides of the translocal family network have, over a period of time, come to understand that telephone calls are the 'new' method of showing care. The more successful instances of sustaining translocal communication tend to be characterized by a certain consistency of routine to call at certain times on fixed days of the week. Routine questions are also repeated even when the answers do not vary much as they represent a routinized way of showing care and intimacy over the phones when co-presence is not possible. By and large, technology has provided a viable alternative to bridge the distance between parent and migrant child, although virtual presence is inadequate in some cases as a substitute of situatedness as discussed below.

In a few cases, and especially where older, widowed men and women are concerned, translocal communication cannot substitute for regular day-to-day care or face-to-face contact between parents and children. Personalized emotional and moral support only has real meaning when expressed in the context of everyday activities rather than words, as the case of Na Ma.[4]

Na Ma (65-year-old farmer) lives alone in a relatively big house in one of the four study villages near Chiang Mai and was happy to host Brian during fieldwork stints. All of her children have grown up, have families of their own, and are living in Chiang Mai and Chiang Rai provinces. Until recently, she lived with her granddaughter (in her twenties) whom she treats as her own daughter as she has been taking care of her for the last 20 years since her son (her granddaughter's father) passed away. Her granddaughter now works as a live-in domestic helper for a Canadian expatriate family in Chiang Mai city which is about two hours away (by car along the highway) from where Na Ma stays. Na Ma chats with her granddaughter every other day or 'whenever my [granddaughter] is free to talk'. She described her relationship with her granddaughter as close given prior investments in caring for her and 'staying in touch' (c.f. Finch and Mason 1990).

However, Na Ma mentioned ever so often that she felt lonely in the house. Clearly, telephone lines were insufficient substitutes in dealing with the lack of

---

4   Pseudonyms are used and the prefix '*Ca*' and '*Na*' in a Lahu name indicates whether they are men or women respectively.

physical co-presence of family members. While Na Ma was watching her usual evening soap opera in the quiet of her living room, Brian asked her if she felt lonely. She replied with a simple '*cao*' or 'yes' in her Northern Thai dialect. Her expression was one of helplessness and to a certain degree, acceptance, knowing that she had little control over the situation and could not do much to assuage these moments of loneliness in her present life. She added, 'It's lonely because no one else is at home and staying alone at home watching TV by yourself is rather lonely. I rather go and sleep if there is nothing to watch'. She mentioned that such forlorn feelings were most intense when evening fell and she went around the empty house locking up for the night. While relations of care are deterritorialized through the phone as seen in Na Ma's communication with her granddaughter, their material and embodied presence within the village remains the preferred relationship of care. Especially when translocal communication is irregular or suspended, and there have not been sufficient efforts invested in keeping in touch on either side, left-behind parents express considerable worry about the lives of their migrant children away from home and the activities they engage in outside of work, albeit with a sense of resignation as seen in the case of Ca Kha (47-year-old farmer):

> *Question*: Do you worry when your son is away? What do you worry about?
> *Ca Kha:* [He has] gone too far away so I don't know anything about him. Worry a lot.... I worry about my son and his life. He goes to work there [the city], maybe [he] follows his friends ... drink alcohol, not do work. Not good.

As seen, left-behind parents' anxieties around being absent from their children's everyday spatio-temporal geographies tend to be expressed in spatialized (e.g. 'gone too far away') and embodied material terms (e.g. 'my son and his life', 'drink alcohol', 'not do work').

Similarly, in the case of Na Yeh (47-year-old farmer who also produces handicraft for a living), it is *both* the spatial and emotional distance from her migrant son (who has hardly been in touch) that troubles her; although in this case, she resigns herself to the sense of translocal disconnectedness by articulating possible resolution of the unhappy situation in terms of respecting her son's privacy and spatial choice, '...if he wants to stay on in the city, he can stay. If he comes back, that'll be good. I will not insist that he comes back to stay with us. *It is his life.*'

Left-behind parents thus deal with the absence of their children in terms of both material changes as well as in terms of the emotions associated with their children's moving. As Glen (1983: 43) observes in a different context, 'Because work and family life were almost entirely separated, intimacy between parents and their children diminished'. When intimacy diminishes, left-behind parents often react by becoming increasingly worried about their children's well-being to the point of questioning the purpose of their migrating in the first place. They question the original rationale for their children's migration to the urban centres and whether the means justify the ends. And for some left-behind parents, they deeply regret the decision they made in allowing their children to migrate for work or in sending them to urban centres to further their education. 'Staying in touch' (Baldassar 2007) hence requires both strategy and 'emotional labour' (Hochschild 1983) – these investments sustain interactions between the nodes in translocal networks, through which acts of care and intimacy can be transmitted. There is yet a limit to the transmission of translocal care and intimacy – the emotional lives of left-behind parents in the translocal village often requires physical co-presence in between periods of translocal (dis)connections – something which can be facilitated by return visits on the part of their migrant children.

Returning to Visit

One weekend, Na Pu's (50-year-old farmer and small-time trader) son, Ca Sor (22-year old university student) returned home to visit. The journey from the university at *Phayao* province to the village by motorcycle took around five to six hours on the highway. Ca Sor had returned home to visit because he missed his mother but also because his uncle had asked him to help out at the Learning Centre as they were conducting a special weekend programme for the children of the village. The evening he came back, Ca Sor's family celebrated by having a special meal together.

Among the four Lahu villages in this research, the frequency of return visits varied across the families as expected, given the wide range of jobs and educational opportunities which the migrant children were engaged or enrolled in. Some migrant children made frequent visits back to the village as often as once a month but most visits occurred once every four to six months, constrained by school calendars, available holidays, the nature of the jobs and financial reasons. Reasons for home visits vary. Writing in the context of Caribbean migrants in Toronto preparing for return migration in the future, Duval (2004: 52; see also Asiedu 2005) argued that migrants make return visits not so much for specific reasons to return, but because return visits 'may instead be the means by which social relationships are maintained'. Among the study villages, return visits were anticipated as opportunities for the migrants to re-establish ties with families and friends in the village, and were occasions marked by special meals and gatherings. Of less relevance to the interviewees was the notion of returning 'home' in order to reclaim a sense of cultural identity that features prominently in the literature on diasporic ties among overseas communities (Glick Schiller and Fouron 2001). Ca Mo, a 21-year-old tertiary student studying in Chiang Mai province who made regular return visits to his village during public holidays and school vacations, indicated that he was not the least bit concerned about his Lahu identity as he felt sure of it. Asked whether he felt the need to reconnect with his Lahu identity on his return visits, Ca Mo's terse reply in no uncertain terms was, 'I am Lahu'.

Return visits provide the channels through which migrant children maintain and refresh social relationships in the translocal village. During their trips home, both Ca Sor and Ca Mo said that they would always set aside time for their families, and visit other friends and relatives in the village. Many also revert to their household responsibilities prior to moving out of the village. For example, during the weekend in the village, Ca Sor was busy cutting the grass around his house, helping his mother, Na Pu, at the farm and by running errands for her, and visiting relatives. It is during these occasions that translocal family members are able to re-establish relationships in concrete ways and create memories which may sustain distance relationships during times apart.

(Re)producing the Local: Conservation, Memory and Surrogates

As a spatial concept, 'locality' needs to be conserved (Appadurai 1995) and in the conservation of locality, nostalgia – as an attachment to the past which helps us make sense of our present lives – plays a major role. As Lowenthal (1975: 5) writes,

> We need the past, in any case, to cope with present landscapes...We see things simultaneously as they are and as we viewed them before; previous experience suffuses all present perception.

Nostalgic memories are often physically contained in material things or invested in the places which feature in everyday life. For example, Wolbert (2001) showed elsewhere how the use of family portraits and photographs acted as temporal references to people's past, thus (re)producing locality visually through people's memories. Locality is hence (re)produced in a stream of social activities which occur in place. Conversely, material expressions of place provide the medium for the reproduction of the local. Hence, the local is (re)produced through social actions and practices and also expressed in the form of material objects and artefacts in place.

In this study, the life of the translocal village continues to revolve around community and work practices (associated with agriculture as the main mode of livelihood) even as new translocal linkages develop. Left-behind parents actively reproduce the local through the adoption of surrogate relationships and re-focusing on locally available community members for emotional support even as translocal linkages with their migrant children are negotiated, as seen in the case of both Ca Han and Ca Nart below. These adopted surrogate relationships make up for the loss of intimacy which the left-behind parents had previously enjoyed with their own children. Ca Han's (53-year-old farmer, teacher and village elder) lives with his wife in the family home; three of his children are married and live on their own while the younger two – a son and a daughter – have migrated to the city for further education. I (Brian) was invited to stay for a family dinner after the interview with Ca Han and to my surprise, a dozen adults and an equivalent number of children joined in the meal that evening. They were a mix of his neighbours, his children's families as well as his own brothers and sisters. It was not a special occasion to warrant a large gathering of sorts, rather it was a 'normal' dinner on a regular day at home in the village. Dinner was served that evening as we all sat under a pavilion-like structure in the middle of the cluster of houses belonging to the people present at the dinner. Conversations over dinner were lively. The missing nephew and niece were frequently referred to, with Ca Han providing news and updates every now and then. Some family members mentioned that at these evening meals, they would often reminisce about the past and the time spent with the migrant children. When Ca Han was asked whether he felt lonely or missed his children, he did not reply directly but pointed to his family and friends around the table and said, '…there are many people here'. By drawing emotional and social support from the local community of relatives and neighbours for quotidian events such as the evening meal, Ca Han was able to improvize a sense of family during these meals while also making present his absent children through including and remembering them in conversation. What we see here is the creation of a surrogate family within physical proximity in order to at least partially reproduce the everyday relations between 'family members' – this is done in such a way as not to exclude the children who are physically away but instead draw them back into the family circle, at least momentarily, through acts of remembering.

Ca Nart (31 years old) – who we referred to at the beginning of this chapter – is the pastor of the church at one of the study villages. His older daughter is currently studying in a town some distance from the village or about three hours' journey on a dirt track and only returns home during the school vacation to spend time with the family. Meanwhile, Ca Nart and his wife devote their time to their church ministry. Ca Nart is perpetually swamped with work throughout the entire week. On weekdays, the church functions as an after school daycare centre for youth and children in the village, hence numerous activities are planned to cater to the interests of this diverse age group. Some of these activities include learning new songs, tutoring and video screenings on the only television set in church. Additionally, Ca Nart devotes his weekday evenings to teaching the children who are now under his care during those evenings how to read and write[5] the Lahu language while Sunday evening language classes are reserved for the adults in the village. Other than occasional meetings with government officials in a neighbouring town to iron out administrative matters – usually on behalf of the uneducated in the village – Ca Nart busies himself with ministry amongst the local village community. The absence of his older daughter in the confines of his home and village has led Ca Nart to look for substitutes to occupy his time, in the hope that he will not be 'too preoccupied' with an absent family member. In his view, the teenagers and children in the village, whom he spends so much time with during ministry and are located in closer proximity than his daughter who is in town, have now become

---

5   The Lahu language does not have a written script but uses romanized English alphabets to translate the spoken Lahu into written form.

his 'children' and the adults, his 'siblings'. Throughout the interview, Ca Nart kept interjecting, 'I stay here not very long but the village is my family!'

For left-behind parents like Ca Han and Ca Nart, localized relationships are hence reproduced through the active adoption of surrogates to make up for the loss of intimacy once enjoyed with their own children. These parents find stability in developing substitute local relationships as they continue to conserve and strengthen ties within the local community.

The migrant child's room in the family home is a place that visually reproduces the local through nostalgia and the invoking of past memories of how the place was once occupied (Lowenthal 1975; Wolbert 2001). The materiality of the place creates a sense of stability in the relationship between the parent and migrant child. Consistently in many homes in the Lahu villages, the migrant child's room although vacated, is conserved in its original form. It is left intact and untouched regardless of the number of years the family member has been away from home either for studies or work. An excerpt from a conversation with Ca See's (42-year-old farmer and small-time trader) family illustrates this.

> *Question:* I am just curious, what is Ca Weet's room like now? Is it still the same life before?
>
> *Ca See [Father]:* Now? It is the same. Never change.
>
> *Ca Sak [Brother]:* Yes the same. Last time our room slept two people but now that he is studying in Bangkok, I am sleeping alone on the bed.
>
> *Question:* How about his things? His personal belongings?
>
> *Ca See:* He has moved most of it to Bangkok. But things that he left behind, we didn't touch because it is his things. So when he comes back, he knows exactly where they are.

Regardless of the number of years their children have been away, left-behind parents like Ca See cling to the glimmer of hope that their children will return home soon. At work here is not just a sense of nostalgia for the past when Ca Weet was at home, there is also a sense of optimism that Ca Weet will return, demonstrated in the way the left-behind family aim to keep his room in the original condition as he left it.

Left-behind parents thus use places associated with their children to recount past memories and experiences of their migrant children living in the house, as well as to look forward to a future when those absent will return. This resonates with Lowenthal's (1975: 7) observation that 'If the character of the place is gone in reality, it remains preserved in the mind's eye of the visitor, formed by historical imagination, untarnished by rude social facts.' The preservation of locality with its associated memories is hence part of family strategy in negotiating change – in this case brought about by the increasing translocal (dis)connectivities with places beyond the village forged by the migration of their children out of the villages – as Chamberlain and Leydesdorff (2004: 231) observed in a different context:

Here, memories of the family (and family memories) play an important part in our perception of ourselves and others, and necessarily are implicated in the negotiations any one individual will make between cultural spheres and in the process of accommodating a new personal stability.

By conserving the room in the original state it was left, with all its memories intact, the left-behind parent attempts to reproduce what was before and create a sense of stability in their relationship with their migrant child. This reinforces their identity as parents responsible for the nurturing and upbringing of their children despite physical absence, and in turn, through reminiscing about the past when their children were present, parents are able to 'know themselves' (Sutton 2004) and be assured of their status in their children's lives. A sense of stability in the present, as well as a sense of hope for the future, is achieved by reproducing the local through retaining and conserving the children's personal belongings and dwelling places. Material expressions of [an unchanged] locality (Appadurai 1995) are hence significant for left-behind parents experiencing new pressures that come with the growing emergence of translocal linkages in the village.

Conclusion

We have argued that the left-behind family in the villages which have experienced outward movement of younger people is not necessarily incapacitated and should not be understood in pathological terms in terms of loss, abandonment and ossification. Social and emotional connections of care and intimacy were mainly sustained through translocal communication between left-behind parents and their migrant children. Families have readapted traditional notions of parenting by incorporating the element of distance in the routinized transmission of care and intimacy through sustained translocal links. However, there is still a place for local co-presence because of the limitations in translocal communication. Attention needs to be given to the local dynamics in particular during periods of translocal disconnection where there may be a suspension of emotional support. Effort and skill have to be invested in 'staying in touch', and sometimes this can only be achieved when the migrant children return to visit. More broadly, what this study demonstrates is the continued significance of the embodied, corporeal, material as well as imagined qualities of family relations as they unfold in everyday rhythms, even as these are being stretched across space beyond the village. This version of translocality, as Brickell and Datta (this volume) explains, takes into account a range of spaces and places without losing site of the embedded, embodied, and material quality of the exchanges between these spaces and places in everyday life.

This chapter has shown that the translocal village is a product of dynamics from below, lending more weight to the notion of transnationality from scales other than at the totalizing assumptions of national space in the transnationality literature. Building on but extending Appadurai's (1995) notions of translocality, we have argued that locality is produced through the constant interaction and struggle amongst different social actors involved in ongoing negotiations as part of everyday life. These translocal villages are situated in specific contexts, and are in an unceasing state of flux, constantly becoming places of a new and different sort. At the same time, they are embedded within a multiplicity of translocal connections and linkages extending to the towns and cities that are shaped by their own history of negotiations and interactions. Translocal influences, in a non-hierarchical manner across multiple spaces extending beyond the village to the towns and cities, have created diversity in the everyday lives of the translocal village communities.

As a countermeasure to provide stability in the lives of the village community against the effects of translocal influences, a sense of the local steeped in past memories is sometimes reproduced. Such a preservation of the local can be seen in the way left-behind family members develop surrogate relationships to take the place of as well as remind them of their absent children. In reproducing the local through the forging of such relationships, surrogates provide a sense of stability and continuity in the lives of the left-behind families. A sense of unchanging locality is also seen in the way left-behind parents conjure up memories and feelings of nostalgia through conserving the private space of the migrant child, and in the process, create a sense of stability for their own identities as parents. As Smith (this volume) has written, making place or the village translocally is a meaning-making practice. The materiality of family relations through the private space of the migrant children, and the situatedness of the family in the village reflects the recuperation of notions of locality. Everyday life in the translocal village hence is a terrain of negotiation between the 'local' and the 'translocal', producing new and nuanced understandings of locality in the process. The 'left-behind' village of the Lahu people is thus an active translocality in the making.

Chapter 4

British Families Moving Home: Translocal Geographies of Return Migration from Singapore

Madeleine E. Hatfield (née Dobson)

The very notion of return migration speaks to the continuing importance of place in the experience of migration and opens this chapter's discussion about translocality and home for returning families. Specific locations are integral to the definition of 'return' because it occurs when migrants go back to a place of origin – somewhere they are from, or they or others perceive them as being from – in which they may, or may not, have lived before. The role that place plays in characterizing this kind of migration also lends it a paradoxical position because the very fact that the location being migrated to is thought to be a familiar one means that the influence of place can be overlooked, both by migrants themselves and by migration studies. This chapter instead draws on research that has centralized the importance of place in exploring the everyday nature of return migration in more depth, especially the small scale domestic places in which migrating families make themselves at home. In its consideration of the value of a translocal approach to investigating experiences of return migration, this chapter also addresses the concept's contribution to understandings of return in the context of migration and migratory pathways more broadly. Perhaps most importantly, the notion of translocality is one way of reconceptualizing separations between different migratory journeys, highlighting connectivity between the day-to-day localities that migrants move from, to and between.

The discussion below begins with an introduction to existing theories of return migration and their relationship to the notion of translocality, particularly highlighting this through place and the everyday as part of the experience of return. Next, I introduce the empirical research this chapter draws on, which comprised in-depth, multi-method research with British families returning to the UK after having lived in Singapore. This empirical research is then used to explore the translocal nature of return migration, especially the role of micro-scale locations and practices in migrant homemaking. It emphasizes in particular the role of homes in the UK as enduring localities for migrants, even when they were physically absent during their time spent living abroad. These homes were symbolically and imaginatively incorporated into the temporary homes migrants made in Singapore. On return, migrants moved back to these specific home spaces, not just to the nation in general. This did not mean an end to their translocality, however, as they then transported souvenirs of their home and everyday lives in Singapore back to the UK. These translocal practices highlight the importance of this scale of analysis by illustrating how domestic spaces are central to migrants' homemaking and experiences of return migration. Domestic homes operate as translocal sites because they are locations to which connections are made and maintained; and because it is within them that these connections and subjectivities are incorporated into migrants' daily lives. I use the term 'domestic home' at times to distinguish house-as-home from wider senses of home, which will be discussed below.

Theories of return migration and translocality

Return migration has been researched much less rigorously than outward migration, despite being a common part of the experience of migration, either in reality or as a 'mythical' intention on the part of migrants (Richmond 1981, King 2000). However, growing research on the topic overturns taken-for-granted assumptions that return entails a straightforward reversion to a life once lived by highlighting the challenge of returning to a location in which one has previously lived or to which one is thought to belong (Hammond 1999, Blunt 2005, Harper 2005, Christou 2006d, Wessendorf 2007). Further, return migration should not be understood as a straightforward process because migrants' mobilities may be variously limited by a number of social, political and economic factors when it comes to return, as on outward migration (Cresswell 2006, Blunt 2007). Research indicates that return might therefore be better understood simply as another migration, to the extent that Hammond (1999) suggests removing the 're' prefix from terms like return and readjust. Theories of cyclical migration and transnationalism have also been applied to the concept of return, drawing attention to it as a migratory journey in its own right, rather than simply the reversal or end of another one (Ley and Kobayashi 2005, Vertovec 2006, King and Newbold 2008).

At the same time, it is important to consider differences in the experience of return compared to outward migration. The particular qualities of return locations and an assumed knowledge of them is an integral part of migrants' connections to these places and often part of their motivation for moving there (e.g. Blunt 2005, Christou 2006d). This assumed familiarity on the part of migrants themselves can, however, make these kinds of migrations more difficult if there is a disjuncture between their expectations and the reality of returning (Martin 1984, Harper 2005, Wessendorf 2007). These expectations are specifically located in places of return such that these types of migrations bring to the fore the continuing role that place plays in experiences of migration, regrounding theories of mass migration, globalization and overarching networks (Smith and Eade 2008).

At its broadest, return is understood as a return to a 'home', most usually a home nation, hence the use of the term repatriation which comes from the Latin *repatriare* meaning 'return to one's *country*' rather than one's home (Soanes and Hawker 2005, my emphasis). This prominence of the nation, however, can overshadow the local, which is the scale at which many migrants experience return. A similar effect is produced by the idea of a 'homeland', which is often associated with the longing of diasporic groups for a real or imagined location that may not be a nation-state (e.g. Brah 1996). As Koser and Black (1999) have highlighted in the context of refugees' movements, a true return to a place of origin is not necessarily achieved by repatriation to a home nation or homeland but by returning to a specific house or piece of land.

Elsewhere, a variety of different return trajectories have been identified, which suggest that migrants choose to 'return' to particular new locations within the nation. This includes 'returning' to major cities and other new locations in preference to smaller settlements of origin where opportunities may be limited (King and Skeldon 2010). In other instances, 'return' is facilitated by investment in new houses in a nation of origin and therefore occurs to these new homes, sometimes in different regions than that of origin (de Haas 2006, Datta 2008). This research, however, is mostly concerned with return migration from more to less economically developed nations and there is little research on how this might operate in different contemporary contexts.

The variety of scales at which return takes place parallels reminders that 'home' can also be experienced and identified at a number of scales from, 'the domestic to the global' (Blunt and Varley 2004: 3). If return is understood as going back 'home' then simply identifying return as occurring on a national scale does not go far enough in addressing the experiences of many return migrants. This does not mean that the nation is not important. As well as being one scale at which 'home' can be identified, it remains both a practical concern in terms of rights of residence, a physical entity as a broader 'homeland' and an emotional or imagined one as the basis for nationalist feelings and attachment. Within this, however, smaller scale locations – or, localities – play an important role in the experience of return, particularly in terms of migrants' day-to-day lives. This is not, of course, to suggest that different scales are easily identified and separated from one another and indeed the concept of scale has been the subject of much debate (see Marston et al. 2005) but it is useful here in differentiating between levels of analysis.

The term translocality offers a way of both conceptualizing and drawing attention to smaller scale places or localities in return migration. Recent uses of translocality have been coupled with contemporary aims in the social sciences to highlight the everyday realities that comprize research subjects' lives. The adoption of such intentions has been advocated in migration studies as helping researchers to explore the 'rootedness of migration in everyday life' (Halfacree and Boyle 1993: 339) and the 'everyday texture of the globalizing places we inhabit' (Conradson and Latham 2005: 228, see also Ho and Hatfield 2010, for an overview). These aims are not exclusive to the notion of 'translocality' and do not in themselves distinguish it from 'transnationality'. Indeed, the quotation from Conradson and Latham (2005) above is taken from an article on 'transnational urbanism'. What translocality offers that is different, is an emphasis on the local, rather than the national. As Hannerz (1996: 6) points out, 'there is a certain irony in the tendency of the term "transnational" to draw attention to what it negates – that is, to the continued significance of the national'. While this means neither that the term lacks utility in understanding the ways in which people make their lives across different places nor that the nation lacks relevance to people's everyday lives, the national is not always the most appropriate – or the only – scale at which to discuss these processes.

The original working title for the research behind this chapter contained the word 'transnational' because of its interest in the way migrants make their lives across multiple locations. Indeed, a number of nationalist and nation-based tropes did emerge from the research, particularly aesthetic notions of homes, gardens and wider landscapes as having particular, often idyllic features (see Daniels 1993). Participants also perceived return as occurring to a home nation, something made even more evident by slight differences amongst two participants who identified as Welsh in a predominantly English group of participants, all of whom were returning to houses in England (this also highlights the complexity involved in identifying national affiliations and places of origin).

Further, the following discussion of translocality must be understood alongside participating migrants' status as 'transnational elites'. In a rare study of British migrants in Singapore, Beaverstock (2002: 527) researched the work and home places of 'transnational elites' similar to the heads of households in my own research and which he defined as, 'the highly educated, highly-skilled, high-paid, highly mobile and "translocal" (Smith 1999) corporate actors/ agents of global capital'. The analysis below explores in more depth the translocality of these transnational elites, which adds to critiques of the notion that such migrants live in a 'frictionless world' (Willis et al. 2002: 505). Despite their transnational relocations, the movement between the homes participants inhabited in different nations operated very much in terms of translocal connections and subjectivities. This provides an important insight into the way that migrants move between domestic homes at the same time as they move between nations, highlighting that these scales operate simultaneously: participants negotiated return in local, including home, work and school, places as well as in nations. These localities are the places in which return to a home nation or homeland is experienced in everyday life.

Much existing research using the term 'translocal' has focused on cities as sub-national places with specific qualities that influence and are influenced by transnational processes (e.g. Smith 2001, Conradson and Latham 2005). Interestingly these places are somewhat larger in scale than the neighbourhoods and villages that were the subject of Appadurai's (1996a) discussion attributed with the genesis of the term 'translocal'. The present chapter considers highly localized places in its discussion of translocality by turning to domestic home spaces as translocal sites. It follows in the vein of research emphasizing the importance of everyday life in migration studies and considers home – and, as part of this, family (see Christensen et al. 2000, Rose 2003, on this relationship) – as integral to the everyday experience of return migration.

In highlighting the importance of homes as physical locations, this research also employs theories around the importance of domestic material culture, which come particularly from anthropology (see for example Miller 1998, 2001). These approaches centralize the physical surroundings and 'things' that affect and are affected by everyday life, with the home identified as an important site of everyday life (Cieraad 1999) as well as the locus of many such 'things' (Miller 2001). As Blunt and Dowling (2006: 23) remind us,

> Home does not simply exist, but is made. Home is a process of creating and understanding forms of dwelling and belonging. This process has both material and imaginative elements. Thus people create home through social and emotional relationships. Home is also materially created – new structures formed, objects used and placed.

The process of homemaking is brought to the fore by migrants because they may be relocating or recreating such material, imaginative and emotional elements of home. Exploring the material aspects of this has the potential to reveal more than just the material itself, although this is also an important agenda in foregrounding translocalities. At the same time,

> … through dwelling upon the more sensual and material qualities of the object, we are able to unpick the more subtle connections with cultural lives and values that are objectified through these forms, in part, because of the particular qualities they possess. (Miller 1998: 9)

These 'qualities', which allow objects to maintain connections beyond their own physicality, combined with a materiality that allows them to be transported (Hooper-Greenhill 2000), mean that they can be valuable homemaking tools on migration. This means that 'home' is not necessarily something left behind on migration or something simply returned to afterwards, but something transportable and always in the process of being formed (Rapport and Dawson 1998).

Consequently, a focus on domestic material cultures can reveal important aspects of the experience of migration, particularly how home is made, understood and negotiated. For example, Tolia-Kelly (2004a) used artefacts as a basis for her exploration of the identity of British Asian women, highlighting both the importance of objects themselves as providing connections to elsewhere and their methodological utility in making these connections evident. Walsh (2006) has also shown how the transport of particular items of material culture can enable migrants to actively transport and (re)create homes in different places. This is in part because of the particular connections objects possess in relation to wider social and cultural life (as suggested above by Miller 1998), connections which can also be highly personalized, as shown below (and see also Morgan and Pritchard

2005). A translocal approach helps to elucidate such negotiations of home and their relationship to experiences of migration, as shown by the discussion below.

Researching with British families returning to the UK from Singapore

This chapter's discussion of translocal geographies of return migration is based on in-depth research with ten British households, which included parents and dependent children or young people (dependency being used to indicate different positions within the household rather than an arbitrary age limit on what constitutes a 'child' or 'young person' – see McKendrick 2001). Working with whole households allowed the research to fully explore the domestic environments in which migrants lived on a day-to-day basis. It also included all migrants as equal agents in the experience of migration, whether the – usually male – 'lead' migrant or 'tied' migrants such as accompanying partners and dependent children (for more on the importance of this see Hardill 1998, Willis and Yeoh 2000, Bailey and Boyle 2004, Dobson 2009).

This does not mean that all individuals within these households experience migration in the same way. Even on the surface, differences were evident in the roles performed by each as all but one participating household had moved as a consequence of the husband/father's job and the majority of wives/mothers were not in paid employment. Households moving together with dependent children were chosen deliberately to enable a consideration of a breadth of different subjectivities within migrant households as a result of these dynamics (each household was also a nuclear family, although this was not a deliberate recruitment policy).

This focus on households in which all individuals migrate together differs from research on 'transnational families', which are defined as those separated by physical distance in order to achieve an economic or social goal. However, this distinction is a blurry one at times because many families moving together are temporarily separated when, for example, children spend periods of time in boarding schools or a parent moves ahead of their family to start a new job. This means that household experiences of translocality are complex ones, involving localities occupied by different people at different times under different circumstances, something which also emphasizes the translocal nature of their migrations.

From the outset, my research was grounded both in nations and in smaller locations within them, being focused on migrants' domestic 'home' spaces in two nations. In order to understand the experience of return more holistically as part of the migration process, I conducted research with three households living in Singapore as well as seven households living back in the UK. On the Singapore side, two of these households were moving back to the UK in the weeks after I met them, providing an insight into migrants' expectations and priorities prior to return. The other Singapore-based household had moved from the UK to Singapore and back, then back to Singapore and intended to return to the UK again in the future (thus offering reflections on the experience of return in both directions).

Households in the UK had been back for various lengths of time from less than one year to eight years.

The nations involved in this research meant a focus on migrants from the UK migrating to and returning from the Republic of Singapore, a nation with a growing community of British citizens who are often on short-term contracts as skilled labour migrants (Beaverstock 2002, Sriskandarajah and Drew 2006). I used defined host and return nations in order to limit the number of different locations influencing migrants, but many participants had also lived elsewhere. This wider context no doubt influenced the practices the research explores, although there are likely to be similarities with migrations between different locations. At the same time, it provides an important counterpoint from research cited above, which is primarily based on return from high to lower income countries.

All participants were middle class and most had migrated on temporary but supportive expatriate contracts, which often included the payment of relocation, accommodation and schooling costs. The short-term and financially beneficial nature of these arrangements meant that many migrants in this position maintained their domestic homes in their country of origin, which were usually let out during their owners' time abroad. This is primarily because of an intention to return and a wish to remain on the property ladder, made feasible by the economic conditions of their migration. Consequently, the homes migrants made in their host country operated alongside homes maintained – both physically and imaginatively – in their home nation. This lends a particularly interesting and poignant aspect to the discussion of translocality below.

In particular, the empirical discussion that follows highlights the importance of participants' homes in the UK, which were symbolically transported to make home in Singapore, and then physically returned to (though not without changes occurring to the occupants, their belongings and the buildings themselves – see Hatfield 2010). On return, these translocal connections were maintained in part through the transport and display of souvenirs, which turned participants' UK homes into translocal sites. This produces and maintains translocal subjectivities (see Conradson and McKay 2007) and places at the scale of individual migrants, the people they live with, the buildings in which they live and the things in them.

This understanding of return as a translocal experience is made apparent via an in-depth methodology, which allowed access to taken-for-granted elements of migrants' lives. This included family focus groups; self-directed photography where participants photographed and captioned the 'things' of importance to them in the home; and follow-up individual interviews using the photographs as the basis for discussion and tours around the home. I reproduce below some of these photographs as well as their captions and verbal quotations from interviews (using pseudonyms to preserve participants' anonymity).

Creative and multi-stage methodologies like this can be helpful when asking research participants to think about practices and objects that they might usually take for granted (Latham 2003). As Conradson and Latham (2005: 228) suggest, there are many such activities that comprize the real lives of migrants but which can be hidden by their ordinary nature:

> Viewed from this quotidian angle, even the most hyper-mobile transnational elites are ordinary: they eat; they sleep; they have families who must be raised, educated and taught a set of values. They have friends to keep up with and relatives to honour. While such lives may be stressful and involve significant levels of dislocation, for those in the midst of these patterns of activity, this effort is arguably simply part of the taken-for-granted texture of daily existence.

One major challenge for research on migrants' everyday lives and the places in which they live them is accessing such taken-for-granted experiences. The method employed in the research discussed below is a focus on material 'things', which helped to elucidate connections that might otherwise have been difficult to talk about and also opened up the physicality of homes and homemaking mentioned in the previous section. Such an in-depth methodology inevitably limited the number of participants, yet as the discussion below will show, it does not follow that a meaningful understanding of migrants' lives must necessarily be limited in turn. Creative and participatory methodologies are also suited to researching with younger research subjects (e.g. Rasmussen 2004, Mizen 2005), making them appropriate for this research, which encompassed children and young people between the ages of five and twenty-five.

Translocality on outward and return migration

Moving home to Singapore

Perhaps the most significant of the findings in this research on return migration is the fact that on returning to the UK from Singapore, all but one participating household had returned, or intended to return, to the same house in which they had previously lived. Existing research suggests that return to a home nation often means moving to a different location than that of origin, either by choice or lack thereof. However, in the context of the research under discussion here, return occurred not only to the same region but almost exclusively to exactly the same domestic dwelling. This may be influenced by the location under study as all households lived in the southeast of England prior to migrating to Singapore, meaning that their former homes were located near the nation's metropole, an attractive location for its economic opportunities. However, other factors also played an important role here, providing more depth to understandings of these localized experiences of transnational return migration.

Economic practicalities were the dominant reason for participants maintaining their homes in the UK and many told me that they had not wished or intended to return to the same house on moving back to the UK (see also Amit-Talai 1998).

However, in some instances, return to the same house was made necessary by other practicalities, particularly returning at short notice such that alternative arrangements for accommodation or schooling could not be made. The difficulties encountered as a result of short-notice returns highlight the fact that, while privileged in their experience of migration, 'transnational elites' such as these are not always in control of their migrations. Interestingly, however, even when such a direct return had been made with initial reluctance, only one household had subsequently moved again within the UK.

These homes in the UK highlight the translocal nature of participants' experiences of migration, which are very much located in domestic home spaces. In Singapore, participants' UK homes were maintained in their ideas of 'home' and this allowed them to live much more transiently abroad, knowing that a more 'permanent' home continued to exist in the UK. Physical distance from these houses obviously means that they did not continue to play the same homely role for participants in terms of being a roof over their heads. However, they did continue to be a home, somewhere to which participants returned and which was not wholly replaced by their day-to-day home in Singapore. When participants later returned to the UK, their return took place not to the nation in general, but to these specific domestic spaces that provided a sense of home.

The multiple homes of migrants have also been identified in other migratory contexts, in which they take different forms (for example, Kenyon 1999, on students' parental, term-time and future homes; and Datta 2008, on the future homes of Polish builders in London). For the migrants in my own research, their next home was not just a future but also a present home. This was a product of socio-economic circumstances that meant they did not have to wait in order to form these homes, which instead already existed and operated as points of continuing security.

For example, the Thomas family, who were living in Singapore at the time of my research, had already returned once to their house in the UK and intended to do so again in the future. Their story illustrates the role that these homes can play for migrants while they are living abroad. Parents Liz and Greg chose to photograph an old, framed black and white photograph of their house in England as important in their home in Singapore. Liz's language was emotive when captioning her photograph with why its subject was important to her: 'An old photo of our house in England which is very dear to my heart'. The security the house brought seemed foremost in Liz's mind when she said, 'it's quite nice to know it's there' and that 'one day we will go back there, I would hate it if I didn't have a home to go back to'. For Liz, the house was not just a building but a 'home'.

Greg's language was more practical, describing the same photograph simply as, 'Our house in England', but he spoke fondly of the surroundings, telling me that there was a cricket pitch outside and that 'you can see all the way to London across the fields'. This is a view he evidently enjoyed and an echo of notions of an idyllic English landscape, one which is both generalized to the scale of the nation but also originally located in the rolling topography Greg described (see

Daniels 1993). While living in Singapore and before their envisaged return, the photograph, which was waiting to be hung on the wall of the family's new house when I visited, provided a daily reminder and presence of this distant home while they continued to make their home in Singapore.

For different individuals within the home, however, this can be experienced in different ways – Liz's language is more sentimental than her husband's and also has a different geography, being focused on the domestic while Greg expanded his attachment to the countryside and view beyond the house. This hints at gendered differences in perceptions of home, with women's being more firmly located in the domestic sphere but many other women in this research also referred to their home's wider location. It also highlights the ways in which localities operate in conjunction with one another so that the domestic can be bound up in its neighbouring environment but also in national ideas of landscape.

The physical nature of the process of display related to this photograph of 'home' is a reminder that such visual images are also 'material beings' (Edwards 2002: 67). Conceiving them as objects, rather than just images, highlights their potential as mementoes due to their portability and the way in which they represent distant times and/or places. Photographs are an established way of freezing time, capturing a specific moment and preserving it (Hirsch 1981, Edwards 1999). They are a particularly valuable memory tool because of their visuality and the fact that, 'recall and remembering requires the utilisation of visual as well as other sensory cues as core elements of the imagination' (Harrison 2004: 33). These characteristics mean that the transport and display of such images in homes in Singapore enabled the transport and display of elements of home from the UK. Photographs' capacity to provide a solid picture of where one has come from could be especially valuable in the context of migration when they, 'provide perhaps even more than usual some illusion of continuity over time and space' (Hirsch 1997: xi). The word 'illusion' points to the deliberate production and preservation of particularly nostalgic images showing places frozen in time when they may no longer resemble their representation. Nostalgia is often identified in migrants' views of their home(land)s and these 'elite' return migrants show how spatiality and temporality intertwine in many migrants' imaginations to produce particular kinds of (trans)localities (see Blunt 2003, on 'productive nostalgia').

These homemaking practices were not limited to adults. Liz and Greg's thirteen year old daughter Amy gave me a different photograph that she had on display in her bedroom. It is a very picturesque, more recent image of the house with a red front door and adorned by hanging baskets, which she captioned as, 'My house in England' (Figure 4.1). Despite using the word 'house' rather than 'home', she later told me that, 'This is just my house [in Singapore], you know, but that's my home'. She said that it was important because it was the only place she had lived in England and that, 'I have so many memories in that house so it'll always be my home, even if we live here…' Despite living in Singapore, then, Amy identified her home as being in the UK and very specifically located in this house. This duality of home in some ways reflects understandings of homelands as long-term sites of belonging particularly in cases of diasporic migration, where a day-to-day home also exists in a host location (Hobsbawm 1991, Brah 1996). In this context of temporary migration, however, the long-term 'home' was also a small scale, domestic space that remained constant and would be – and indeed already had been – returned to.

Figure 4.1 'My house in England' [Amy]

Moving home to the UK

Amy's reiteration of some of her parents' attachment to their home in England also draws attention to the intertwining of translocality, home and family. In the context of migration, these relationships have important consequences. Transnational families, for example, can find that 'home', as well as family, becomes fragmented as a result of separation from kin (see Parreñas 2001). Conversely, when families migrate together, they may find that they are moving 'home' with them (see Hatfield 2010).

The taken-for-granted nature of the relationship between home and family can mean that it is easily overlooked but it is important to stress the importance of the physical locations in which these relationships are created, maintained and developed. On returning to the UK, parents and children experienced their move home as individuals within households whose members each had different needs and priorities as well as varied knowledge of the homes to which they were returning. Parents strove to provide the right kind of home for their children, one that offered the family the kind of lifestyle they would wish, decisions very much influenced by their children's needs and desires (and later actively by children's homemaking within these spaces – see Hatfield 2010) but also mediated by the circumstances of their return. These returning migrants had very detailed knowledge of the homes they were returning to as a result of their previous residence in them. This meant that the negotiation of their return was considered at a smaller scale even than a need to find and move to a new domestic space.

For example, when looking ahead to their impending move back to the UK, the parents in the Dawson household spoke with enthusiasm of returning to their open plan kitchen-living area, a type of arrangement uncommon to British migrants' accommodation in Singapore. My visit, though, sparked a discussion between dad Gavin and youngest daughter Erin, eight years old at the time: plans to extend this communal area of the house on return by means of a conservatory had obviously been discussed much earlier but Erin took my visit as an opportunity to try to secure her dad's promise that this would not be built, as it would take the space out of the garden that had previously been used as a play area. Erin's agency in this discussion was in part a product of her knowledge of the house to which she was returning, an agency that had to be negotiated by her parents in their efforts to improve their home for all of its returning residents.

Elsewhere in Singapore, shortly before their return to the UK the Black family were packing up and organizing the transportation of their household from Singapore to their flat in London. Zoe and Duncan had left the UK eight years previously as a couple, and were to return with two sons, seven and five years old. As well as challenging the notion of 'return' for young British citizens who have never lived there (see Christou 2006d, Wessendorf 2007, on returning second generation migrants), these circumstances highlight the localized negotiations of places that occur on return migration. The spacious house the Black household had occupied for six years in Singapore was much larger than the two-bedroom flat they were to 'return' to in London. When showing me his favourite pictures and souvenirs in his bedroom in Singapore, eldest son, Damien, suddenly paused to ask his mum if the large dresser in his bedroom would fit in his new bedroom in London. Since he had never actually lived there himself, he did not know what his room would be like. Zoe replied that it would have to go in the garage, and I wondered how her two sons would adjust to their new accommodation and what would happen to the antique British furniture and large souvenirs from the family's travels in Asia, which were clearly much too sizable and numerous to be part of their domestic space in the UK.

This movement of home back from Singapore to the UK is another element of translocality that is very much intertwined with, but distinct from, moving home from the UK to Singapore. While in Singapore mementoes and memories of homes in the UK played a part in homemaking, many returning migrants moved furniture and souvenirs with them on their return, thereby transporting elements of home in Singapore to the UK. One parent participant, Andrea, for example, told me that she had done a lot of shopping before leaving Singapore, buying things

Figure 4.2 'This Buddah [sic] head is probably the most treasured possession from Singapore.' [Andrea]

that she would have bought anyway but which was made more 'focused' in the face of return to the UK. On returning to the UK, Andrea had deliberately confined most of her Singaporean souvenirs to the dining and bedrooms, where she had tried to create, 'more of the feel we had in the apartment in Singapore'. Some things had been accumulated for general aesthetic or memorial reasons, while others were chosen to fit a deliberate place within the home Andrea knew she would return to. The stand on which her Buddha head is placed in Figure 4.2 had a partner on the other side of the chimney breast, both deliberately chosen to fit these awkward spaces on return to the house. On the other stand was a 'very classic Singapore orchid', the Republic's national flower and a reminder of the botanic gardens next door to where Andrea and her family had lived in Singapore. The domestic arrangement of these objects in their new home illustrates the highly localized practices undertaken by return migrants who meld the architecture of British homes, embodied in this case by the chimney breast of a Victorian terraced house, with souvenirs of the tropics and Southeast Asia.

Andrea was not the only participant to highlight instances of this and indeed the actual objects she chose – Buddha statues and orchids – were often selected by other participants as particularly symbolic of these endeavours. These objects were not always bought explicitly for return and many had played a part in home life in Singapore. Andrea hinted at this when she talked about objects recreating the 'feel' of her home in Singapore and orchids reminding her of the immediate surroundings of this home.

In the Evans family, Sandra also talked about orchids, which, she said, 'remind me of the beautiful tropical plants in Singapore' and the fact that, 'we had them growing in our garden like daffodils'. The Evanses' Buddha statue had also been a prominent feature in their living room in Singapore where they held a number of parties and, on being relocated to their dining room in the UK, acted as a reminder of these fun times. Deploying these objects as symbolic – or metonymic (Gordon 1986) – reminders of Singapore suggests the continuation of return migrants' connections to the things that surrounded them in their everyday lives while living abroad. Pete Evans captured this two-way process when he said, 'There are things that remind[ed] me of home in Singapore and now there are things that remind me of Singapore at home'. This also reflects the relationship between objects, the domestic and wider localities, emphasizing Miller's (2001: 1) assertion that, 'It is the material culture within our home that appears as both our appropriation of the larger world and often as the representation of that world within our private domain'.

Many qualities attributed to Buddha statues and orchids by participants included rather generalized notions of peace, tranquillity and exoticism, which suggest the appropriation of larger-scale, quite essentialist notions of the 'Orient' (see Saïd 2003), extending beyond Singapore itself to the Southeast Asian region more widely. At the same time, however, it is important to recognize that these objects also operated at another level, being overlaid by different – often smaller-scale and more personalized – meanings connected to participants' day-to-day lives (see Morgan and Pritchard 2005). Andrea, for example, captioned her photograph in Figure 4.2 as below:

> This Buddah [sic] head is probably the most treasured possession from Singapore.
> It was given to me as a gift from my sister and her family as a thank you for the

holiday they spent with us. I love the serenity of the face and it symbolizes all I love about Asia.

For Andrea, then, her Buddha head was not just about 'serenity' and 'Asia' but also a reminder of her extended family staying in her home in Singapore; while for the Evans family, a Buddha statue simultaneously represented 'peace and fun' (Sandra Evans) because of its former location in the living room where they socialized. Orchids were not just symbols of Singapore as a whole nation, but reminders of homes' gardens and immediate surroundings. These individualized connections often related to micro-scale experiences and localized spaces, including domestic ones. On return, they memorialized these specific places, people and activities, bringing them into returned-to home spaces as a result of objects' capacity to transport such attachments through their associated meanings and values (Parkin 1999, Hooper-Greenhill 2000, Tolia-Kelly 2004a). These objects maintained connections not just transnationally, but between, of and within specific localities, making translocal spaces of domestic homes.

Conclusions

The lens of translocality used above highlights the emplaced nature of return migration, in particular the ways in which it is experienced in – and in relation to – the domestic homes inhabited by migrants. Such grounding of research on return migration in localities aims to fill in some blanks left by nation-based understandings of what it means to be a return migrant. The case studies presented here show that while return takes place transnationally, it is simultaneously experienced translocally. It is in these localities that migrants live on a day-to-day basis and it is here that they realize transnational connections and subjectivities. This has implications for migration in general as well as – most significantly in the context of this research – return in particular. As an understudied part of the experience of migration, it is important to draw attention to the localized, everyday nature of a type of movement often equated with nations, rather than localities.

The focus above is on domestic homes, which are of course not the only locations comprizing migrants' – and non-migrants' – daily lives, which see the interaction of home, school, work and social places among others. It is through a focus on localities that the nature of migrants' everyday experiences becomes more apparent but also that the importance of these localities can be explored. What a focus on homes allows in particular is an insight into the relationship between return migration and broader ideas of home as a place of belonging as well as a place in which one lives. In the case of the research discussed above, 'home' and 'the home' (Wise 2000: 300) intertwine, with the latter a translocal site in which the former is experienced, maintained and transformed.

On returning 'home', the migrants who are the subject of this chapter moved back to their domestic homes. These homes operated as sites of translocality in two key ways, which influence – and are influenced by – wider notions of belonging: firstly, they are translocal sites *to* which connections are made and maintained through homemaking practices; and secondly, they are sites *within* which translocal connections are made and maintained. Objects, activities and memories transported from one nation to another found their home in these domestic locations. In the examples cited above, in Singapore this meant photographs of homes in the UK hung on the walls of those in Singapore. In the UK, translocal connections to Singapore were facilitated by the presence of Buddha statues and orchids, reminders of daily life in and around homes in Singapore.

Home and return operate across many levels and translocal subjectivities are informed not only by national (or even larger) scale identifications – landscape, exoticism and national symbols among others – but also by neighbour(hood)s, urban areas and regions – the friends at a party, the gardens nearby, the view from the house. Rather than close off these scales, however, a focus on translocalities and translocalism highlights the way all are negotiated in the places inhabited by migrants in their day-to-day lives.

The context of this research is of course specific, again at a number of scales. The migrants under discussion are 'transnational elites' whose conditions of migration allowed (and in some instances required) them to maintain and return to the same homes as they previously lived in, though this may not be the case for all such migrants and may also happen to those migrating and returning under different conditions. The examples of those things in and around the home that became translocalized are also particular both to the nations and the specific locations migrants moved between. They would be different for those moving between different places, as well as those migrating between the same locations but with the 'outward' and 'return' directions reversed. There is no doubt that the translocal connections maintained with the home/return location are different from those with the host location, the former being represented by a long-term enduring connection to a home that may be returned to; the latter by symbols of a life migrants may not return to but which has a place in their lives and homes on return. This, though, is yet another example of the importance of places and the different connections people have to them, connections which, this chapter has suggested, are best explored through localities. Focusing on homes as sites of translocality for return(ing) migrants both grounds the experiences of 'transnational elite' migrants in the everyday, localized places in which they live their day-to-day lives and reveals the nature of homes as translocal sites.

Acknowledgements

Thanks to all participants who made this research possible and to Katherine Brickell, Ayona Datta and Katie Willis for their helpful comments on drafts of this chapter. This research was supported by an Economic and Social Research Council (UK) Studentship.

PART 3:
Translocal Neighbourhoods

Chapter 5

Translocal Geographies of London: Belonging and 'Otherness' among Polish Migrants after 2004

Ayona Datta

It was really convenient. You could go out in your slippers, put the rubbish there, and of course in terms of shopping. We lived just above the off-license. So quick shopping was no problem. So you didn't have to go somewhere far, but you could just go downstairs. To a Turkish shop. And buy everything. But most of all he knew us well, the shopkeeper. Sometimes we even got discounts. ... Lots of Polish people lived there. And of course most of these Turks, these Turkish people who knew me. At the baker's everyone knew me. I always used to buy there. Just around my house, along the main street, in Green Lanes everyone knew me. So, when they saw me, they even stopped me in the street to have a chat, or joke, always. They just treated me, I don't know, as if I met some friend in my own neighbourhood in Poland. The same atmosphere, similar. [Dawid]

Dawid, a 28 year old Polish migrant who had arrived in London after the European Union expansion in 2004 had chosen to live in a Turkish/Kurdish neighbourhood – *Green Lanes* in North London. Although he missed his wife and children who were left behind in Białystok, Poland, he felt that living in a Turkish neighbourhood gave him a sense of familiarity – one which was similar to the sense of belonging in his own town. Green Lanes, Dawid noted was a place in London where he was able to develop a sense of home.

Dawid's narrative reflects what Ma has described as 'the dynamics between localized life worlds in faraway sites' (2002: 133). Dawid highlights a crucial aspect of the lives of Polish migrants in London. He suggests that in London, different urban spaces and neighbourhoods are negotiated through migrants' personal histories, memories and a spatialized politics of difference. His narrative highlights the importance of thinking about the city not simply as a destination for migrants; rather as a space that is fractured through migrants' geographies of movement across different spaces and places. Further and crucial to this narrative is the negotiation of home and belonging in different neighbourhoods of London, through encounters with 'others' who might also be migrants like oneself.

In this chapter I am interested in examining how everyday practices of migrants construct the city through a variety of localized neighbourhoods that are connected to a range of other localities within and beyond the nation. I take 'local' here to be a spatial construct that is produced from its relation to a variety of other spaces and places. In other words, I see 'translocality' as a process that 'situates' diverse spaces and practices within different locales. The 'local' therefore is not simply a place but also a map that people arrive at through their many 'scaling practices' across different spaces (Slater and Ariztia-Larrain 2009). From this perspective, Dawid's sense of belonging to Green Lanes is not only negotiated through global-local connections or national boundaries; but also through a range of local-local attachments along his migratory route.

I focus here on the more 'traditional' spaces of neighbourhoods that are evoked in Dawid's narrative, in order to highlight how these neighbourhoods are also part of the translocal geographies of movement, which connects to other homes, neighbourhoods, cities, regions, and national spaces in very specific ways. I argue that although there has been substantial scholarship on the situatedness of transnational networks and movements among migrants, there has been relatively less focus on the everyday lives of migrants which are often mundane and ordinary. This is not to say that migrants' lives have not been important in transnational research; rather that their lives have been examined largely through economic, political and social networks between sending and receiving contexts. Under this kind of scrutiny, what has emerged is that migrants' notion of home and belonging is multi-faceted and multiscalar, negotiated in the in-between spaces of migration that produces forms of 'transnationalism from between' (Smith and Guarnizo 1998).

Such explorations however are less explicit about the everyday spatial practices of migrants in the city where they enact mundane activities of working, eating, sleeping, walking, socializing and so on. These practices are tied to negotiations within localized contexts of neighbourhoods, streets, marketplaces, and public transport, which are also related to economic opportunities, subjective identities, and social networks in very different ways. Such spatial practices I argue fracture the notion of the city as *the* site of experience into a range of simultaneous sites or locales (neighbourhoods, streets, parks, houses and so on) defined by their specific social, material, cultural, political and geographic connections to other such locales. This produces a translocal city – since the city itself is experienced as a set of interconnected spaces and places that are material, social, and embodied during migrants' heightened mobility across its neighbourhoods. The translocal city becomes a relational experience negotiated between and across different urban sites, which draw upon transnational imaginations and translocal journeys of migrants themselves.

Neighbourhoods as ethnoscapes?

The notion of the neighbourhood has been subject to much critique within and outside anthropology. Appadurai (1996a, 1996b, 2005) as one of the strongest critics of neighbourhood as a bounded territorial form, sees it instead as connected to contextual and relational references. In his notion locality and neighbourhoods are both processes rather than products, and the spatial production of locality is seen to require not just the production of particular material forms but also a fundamental reorganization of space and time. I consider these connections between locality and neighbourhood as key to the construction of a translocal city since it is through the relational construction of particular locales that migrants attempt to make sense of their movements across different neighbourhoods.

But while Appadurai constructs locality as deterritorialized, diasporic and transnational, he critiques the often territorialized and bounded understanding of neighbourhoods as physical and material forms. In other words, Appadurai bases his critique of neighbourhood as relationally opposite to locality and hence a space where the state reigns supreme. Appadurai equates this aspect of neighbourhoods to *ethnoscapes* – a spatialization of migrant identity where their practices produce and reproduce social forms that belong to particular nation states. Ethnoscapes are materially realized in ethnic neighbourhoods, where urban warfare is waged between the migrants and the state. This pessimistic portrayal of the production of physical neighbourhoods constructs migrants as either vulnerable and marginalized or insurgent and resistant in the local context. And it also constructs the state as the primary regulator of agency, identity and experience during migration. Appadurai thereby suggests that neighbourhoods are also becoming deterritorialized, disembedded and post-national. In light of the state regulation over social formations of neighbourhoods, he notes that diasporic populations are creating new mediated formations of virtual neighbourhoods which circumvent state control.

I see Appadurai's detrerritorialized view of neighbourhoods as inadvertently silencing those everyday places which are part of the local-local linkages of migrants' lives. His notion of a post-national and deterritorialized local that is characterized by increased flows and mediatized connections does not capture the often material and embodied nature of these interactions as they are reworked in particular places. My argument therefore, while agreeing with the relational construction of locality and neighbourhood, also departs from Appadurai's distinction between them in a few respects. I believe that the neighbourhood can be realized in a variety of ways that is not necessarily through state control; rather a day-to-day mundane negotiation of the particular localized opportunities that can evoke notions of home and belonging. I see the neighbourhood as a particular material formation of the local but one that is not necessarily territorially confined or controlled by the state – rather constructed during its dialectical relationship to other spaces and places at different scales. The local in this context is a spatial formation which is constructed during particular interactions within material places of neighbourhoods and can exist simultaneously within other scales. Taken together, the locality and the neighbourhood are only realized in a dialectic relation with each other, they are both 'situated' in specific ways and they are both sites where the 'mobile' identities of migrants are emplaced.

Translocal city as a spatialization of habitus

One way to understand this dialectic would be through a spatialization of Bourdieu's notion of the 'habitus' (Bourdieu 2002). Habitus provides the fields of meaning to different forms of (social, cultural, economic and symbolic) capital. In order to operate effectively in this field, actors must learn how to exchange between different modes of capital in different fields, and operate effectively through everyday practices. For Bourdieu, human actors understand their immediate conditions through embodied and affective experiences and negotiate these through particular sets of meanings relevant to the field of power that they are situated in. Considered thus, the habitus can be extended to a spatial realm in order to reconceptualize both locality and neighbourhoods as sites where forms and degrees of social capital are translated, exchanged and reworked.

I use habitus as a heuristic tool to highlight the 'multi-stranded connections' (Kelly and Lusis 2005: 831) between different locales and neighbourhoods. Through a study of Polish construction workers who arrived in London around the time of the EU expansion in 2004, I suggest that these connections give value and meanings to different forms of (social, cultural or symbolic) capital, and reflect migrants' 'situatedness' within particular places in the city. In my other writings around this topic, I have used habitus to situate participants' interactions with and attitudes to 'others' in a number of everyday places. I proposed that participants relationships with different places 'are shaped by individual biographies, access to forms of capital, and the localised spatial contexts in which specific attitudes and behaviours towards others are practised' (Datta 2009a: 367). Different spatial contexts operationalized differential access to social and cultural capital which allowed participants to reflect upon their own transnational histories, ethno-national identities, and access to power in these places (Datta 2008). Everyday places such as building sites, pubs or shared houses were located within wider networks of symbolic capital and social power but were understood through the embodied and corporeal nature of migrants' experiences within these places (Datta 2009b, Datta and Brickell 2009). I suggested that the situatedness of participants in everyday places were crucial to the understanding of their mobility as transnational migrants.

My notion of the translocal city builds upon my earlier discussions to focus specifically on the translocal negotiations among this group of Polish migrants who live in London and work as builders in the home-refurbishment sector. Often transnational migrants are perceived to move to a city and live in its ethnic neighbourhoods, sending remittances home and somehow also changing these places by building churches, houses, markets and even entire neighbourhoods. Largely absent in our understanding are the reasons why particular cities are chosen over others and the choices that re made around living and working in particular urban neighbourhoods. These decisions evoke the different material, embodied, corporeal and affective ways in which participants situate themselves within urban sites, even as this situatedness is premised upon their connections within other spaces, places, times and scales.

I examine the 'translocal city' through the transnational and translocal mobilities of my Polish participants (between the ages of 24 and 47) who were interviewed during 2006-2007. I used a combination of semi-structured interviewing and participant photographs, which I call visual narratives. There were two stages in this – an initial information gathering interview with participants when apart from other factual information we also drew a timeline of their employment and accommodation since they first arrived in London. After the first interview they were provided with a disposable camera to take pictures of their 'life in London'. In the second interview, their photographs were used to solicit the contexts and meanings of their experiences. Participants' photographs reflected a variety of places– homes, parks, public transport, building sites – in short, sites that made up their translocal geographies in London.

While there is a history of migration from East-European countries since the Second World War (Sword 1996), the EU accession as it was called has resulted in the most significant impact on migration into the UK in recent years. In May 2004, nationals from eight East-European states (called A8 states) were given rights to live in and enter into employment in the UK to the European Union (EU).[1] At its peak during 2005 to early 2007, 630,000 individuals registered with the WRS (Home Office 2009).[2] Nearly two thirds (65%) of those registered were from Poland, followed by Lithuania 10% and Slovakia 10%. These workers were male (58%), young (82% between 18-34 years) and without dependants (93%) in the UK (Home Office 2007). Although the number registering have steadily fallen with the economic downturn since late 2007, London still has the second largest concentration (15%) of WRS registered workers in the UK (Home Office 2009).

The notion of the translocal city then takes on the linkages between locality and mobility made by my participants in a number of ways. First, I argue that while transnational mobility brings participants to London, local experiences of mobility within the spaces of public transport, streets, and parks transforms the city into a range of urban sites which are constructed in relation to other localities and neighbourhoods within and beyond the city. I take this as the condition of translocality – although they show heightened mobility across national and local spaces, their experiences within these urban sites are simultaneously highly emplaced. If translocality is a node that pertains to how people move across places (Mandaville 1999), then the city is the site *par excellence* of this translocality, since it involves constant movement of migrants as they attempt to negotiate and improve their everyday conditions at home and work – economically, socially and spatially.

---

1 These rights were premised on the fact that they registered with the Worker's Registration Scheme (WRS) which required all A8 nationals entering to work in the UK to register with the UK Home Office.

2 But DEFRA (2005) found that the number of eligible workers who had not enrolled was close to 15%. As a result, the WRS records underestimate the numbers of A8 nationals.

Following on from the above argument, I suggest that locality and neighbourhood are reworked in the city through highly subjective and material experiences of migrants. These experiences are simultaneously transnational, regional, and translocal. For example in the case of participants who cross the borders from Poland to UK, their identities are increasingly shaped through the regional space of the European Union, which is as much political, geographical, and structural – and which guide the possibilities of 'free movement', work and dwelling. Mobility is both transnational and translocal, across national boundaries and across urban neighbourhoods, with the links that migrants create between different locales in the city and other places far away becoming opportunities for retrieving a sense of agency, home, and belonging in new places. Not all of these links are positive. Indeed as many participants suggest, these links often provide reminders of the betrayal of their dream of migration to London and more generally to the 'West'.

Translocal geographies of London

London as a city is of particular importance in understanding the translocal geographies of Polish migrants in my study. In the socialist state London occupied particular imaginations which were related to the forms of access that participants had to the 'West'. Until the fall of the Communism in 1989, most participants had never travelled to the 'West'. A few older participants had travelled within the Soviet Bloc as tourists, but for them these countries had evoked similar experiences as part of a similar political system. Travel to countries outside the Soviet bloc was restricted and denied to most of them before 1989. In this context, London had been the 'city of dreams', which had been denied to them in the Socialist state. For a long time, London had been the centre of England which had produced the Beatles, Gerald Durrell, and Shakespeare. This was the London which during many years of Communism had been the cultural symbol of the 'West'. Participants would speak of how they would illegally tune into BBC World radio in order to hear the news from the West, which was 'forbidden' during the socialist regime. Coming to London therefore was not simply for work but also to become part of the West.

> To be honest, I was interested in this culture which, which used to flood Poland. All these English language products from the West. Music, films, it was all from here. Or from America. But generally, it was about English. So also that interest in this culture. And, the money is best here, let's be frank. In the whole Europe you can make the best money here. [Andrzej]

Thus even before arriving in London, the city was already constructed in relation to participants' personal experiences of state socialism. After the fall of communism, many participants attempted to come to the UK to work illegally, primarily by overstaying on their tourist visa. Their journey into London was fraught with difficulties – participants arrived via land and sea and were repeatedly turned back if they could not convince immigration officers of their intention to return. They were often robbed along the way or cheated by their contacts. But most participants noted that their first port of call after arriving in London would be Abbey Road – the studios of the Beatles. They took time off during the weekends to visit the many museums and tourist sites in London. In short, when participants arrived they primarily attempted to negotiate their relationships with London as a city of the West – of modernity, high culture and democracy.

Participants therefore arrived in the UK as economic migrants between 1996 and 2006. They were so-called 'circular migrants' – many had arrived and stayed on illegally before 2004, many returned to Poland just before the EU expansion and then arrived in London as legal workers after 2004. Ryszard, a 32 year old Polish bricklayer, saw his 'circular' mobility as part of an adventure, as a process of personal development. Once in London, Ryszard began working as a bricklayer, but he also began studying for an undergraduate degree in Archaeology. Fieldtrips on this degree took him across UK and Western Europe. Ryszard saw this as an essential part of living and of connecting to places – his mobility shaped by a gendered desire to understand other cultures and associate with other places.

> If you are in a mood to discover something, you want to learn something you know, develop as a man. Like it was, I don't know, if it is Polish problem or not. Maybe it's just universal, but er travel gives you education. It's literally, well it should be like experience, it like educates you. Travel educates you, and it's like with you, you notice things and find out things you've never heard before. Or just you can go to a place you never heard or just, just, just knew that place as a description on a map or something. It's lovely; I like that sort of adventure. [Ryszard]

Another participant Karol, a 32 year old Polish carpenter made different connections to London through his personal relationships. Karol had initially come to London in 1996 as a student to learn English, but he married an English woman soon afterwards and moved to Bath (west of England). Karol however, always felt 'foreign' in Bath. 'I loved the house, I was happy with my wife there, but I hated the place. I could not stand it' he said. Within four years, his marriage broke up, and Karol returned to London. He then went to Sydney where his sister lived, took up odd jobs there, bought a car and travelled across the country for a year.

**Figure 5.1 Karol's picture[3] of his car rear windscreen in Australia, in which
he travelled across the country**

During his travels across Australia, Karol began to collect stickers of the
places he visited and stuck them to the rear windscreen of his car. Before leaving
Australia, he took a picture of this rear windscreen, filling in the place of 'Darwin'
whose sticker had fallen off. While there was an inherent touristic purpose on
this image, Karol says that he also wanted to document his relationship with
Australia and record his fond memories of a country which he very much wanted
to live in. He returned to London after 2004 and found a job as a carpenter. At the
time of interview, Karol was living with his Australian girlfriend. His narrative
below suggests that he had formed different kinds of situated subjectivities within
different national spaces; his mobility shaped by wider structural constraints and
opportunities of immigration laws, and his identity shaped by his associations with
different places.

---

3   Here instead of a picture generated during this research, Karol used a picture which
he already possessed. This was because Karol was an avid photographer and he saw the use
of a disposable camera as an insult to his skills. In our interactions therefore Karol used
a collection of his own digital photographs to talk about his experiences of migration and
mobility.

Australia is a country I would love to live in. It's not that easy to immigrate and I couldn't even bother immigrating there. I've changed country once, that's enough. I'm happy and I like London, but Australia is just a place that I would really feel comfortable. ... I'm Australian by heart, English by where I live, and Polish by birth. [Karol]

These narratives suggest the making of a variety of situated positionalities during transnational mobility. Participants make sense of the different places that they travel through by describing themselves relationally to these places and by associating these places with particular types of symbolic or cultural capital. But they also suggest that London as a city became more than just a 'place on the map' as Ryszard puts it. London was already constructed through their experiences in a socialist state, the restraints over their transnational mobilities and personal histories. Before they arrived in the UK, they had already situated London on the crossroads of culture, modernity and western capitalism.

Moving between home and work

This city of their dreams turned out to be quite different when participants began living and working here. Very soon after they arrived in London, participants realized that finding work was the biggest difficulty, irrespective of whether they arrived before or after 2004. These participants came from small industrial towns and cities (Cieszyn, Tarnow, Lublin, Torun, Nowa Huta among others), where they were working in a range of blue-collar jobs in factories and industries, and in a couple of cases as farmers. They had very little language skills and no recognized professional or vocational skills suitable for getting employment in London.

Participants would often hear of a place called *sciana placzu* (Wailing Wall). It was actually a small Indian cash and carry shop in west London, which since the 90s had put up job notices in Polish language on its shop window. Most of this was for casual work – as labourers, cleaners, barmen and so on and hence were largely unregulated. This allowed those arriving without language skills and often illegally to find undocumented work and thus remained as the first destination for job searches even after the accession in 2004. The location was not accidental – the Polish Centre of Art and Culture (*Polski Osrodek Sztuki i Kultury* or POSK) was right next door. Most migrants came with little money, they had to find an income really quickly, but the competition was high, and they did not always have the right papers. As a reflection of the despair and uncertainties accompanying the aftermath of migration in London the Wailing Wall[4] was seen by some participants as a caricature and by others as an exaggerated joke. But none of them denied its

---

4 The Wailing Wall is synonymous to the Western Wall inside the holy city of Jerusalem. The wall is believed to be a part of the Second Temple destroyed in 70AD by the Romans. It signifies a Jewish place of prayer and lamentation.

connections to their personal histories, their initial lack of social capital in the city, and the structural transformations in their political systems and of the European Union that brought them into London in the first place.

POSK on the other hand, despite being a place of 'Polish culture' was not where participants found themselves to be particularly comfortable. Part of this was due to the class differences between the earlier Polish émigrés who had set up POSK as a celebration of Polish high culture. POSK hosts regular films, theatre productions and talks, includes a bookshop and cafe where one can browse Polish books and eat Polish food. It is also located in west London in an upmarket neighbourhood, where the earlier Polish immigrants had settled since the War. But, participants often found themselves embarrassed to enter its premises because of the particular sector they worked in, where their 'ungroomed' look or even paint-stained clothes immediately marked them as out of place inside the building. All the participants agreed therefore that despite being so close to the *sciana placzu* where they went frequently, POSK was not a part their everyday places in London.

Without recognized skills or language in London, participants were recruited in a range of odd jobs in the catering, cleaning and building sectors. They switched jobs frequently and finally began to work as labourers in the building sites because they saw themselves as provided with a larger range of opportunities of progressing up the ladder in the building sector (Datta and Brickell 2009). These building sites were small-scale home refurbishment projects where participants were employed without contracts and paid cash-in-hand often below minimum wages. As temporary places of work with restively fast turnover periods, these building sites continuously shifted across the city. When employed in these building sites, participants had to learn to find their way to these places using public transport. Participants were usually also employed in two of three different building sites at the same time, and since these were often spread across the city, they had to move everyday or every other day to different neighbourhoods in the city where they were building. Thus they became highly mobile actors within the city, continually travelling between building sites as they attempted to remain employed and make the best of the booming construction sector during 2006-2007.

Participants' choices of accommodation also became tied into the spatio-temporalities of building sites across the city. They moved homes in the city as they attempted to live close to these building sites and reduce their commuting distances. This was particularly difficult since these building sites were located in neighbourhoods where rents were high. Participants in turn attempted to live in neighbourhoods connected by public transport. This is consistent with a study by Oxford University (Spencer et. al. 2007), where 30 percent of their East European respondents (which included builders) cited employment as a reason for moving accommodation. This is not to suggest that their residential moves were solely and directly connected to employment; rather that the spatio-temporality of their work formed an important part of their decisions to stay in temporary and shared accommodation without formal contracts, from where they could move out at short notices. The urban neighbourhoods then became important sites of negotiation of migrant experience as they attempted to find not just affordable accommodation, but also places where they could be connected to workplaces, while simultaneously cultivating a sense of belonging, familiarity, and comfort in these neighbourhoods.

Participants thus lived primarily in East and South London neighbourhoods since these had cheaper rents, in shared terraced houses with relatives or friends, or in council flats that were often sublet to them illegally. At the time of interviews they had made between one and seven house moves since arriving in London. Those single in London shared bedrooms with their friends or relatives, while those with partners and children, shared larger terraced or detached homes with friends – in all cases they had informal rental agreements with their landlords. The economic nature of their migration meant that participants were willing to experience the inconveniences of shared accommodation in London in order to keep costs low. Further, the high rents and unaffordable property prices in London made this the only option since they were not usually eligible for social housing in the UK[5]. Their experiences of everyday life in London were therefore shaped by their continuous movement between different temporary 'homes' as they attempted to keep rents affordable while remaining close to the building sites. These movements, not just between Poland and UK but also within rental housing in London, and between the homes that they built in the city, shaped how the city was experienced and negotiated, and the ways that they reworked their own mobilities within the city, within the EU, and across other countries.

Benedykt was one such participant who arrived as a tourist from Torun via ferry into Dover in 1999, but was deported because he could not convince the immigration officers of his intention to return to Poland. A few months later, he flew into Heathrow and entered UK as a tourist. A Polish acquaintance took his money to 'arrange' a work permit for him but was never seen again. Without any money or accommodation Benedykt spent three nights sleeping rough in a park in Barking (east London); from there he would walk everyday to west London to check out new job openings on the *sciana placzu*. He eventually found temporary employment as a labourer in building sites across London, and had to walk up to six hours each day to reach work. During this time he constantly moved accommodation, sharing rooms with strangers in large houses, renting illegally sublet rooms from social housing tenants, and sleeping rough in London's parks, until he was hired more permanently as a labourer by a Ukrainian building contractor. His Ukranian boss provided him with a single room in a shared house, and he brought his wife and children over from Poland at that point. But his wife began an affair with his boss and subsequently divorced him. Benedykt therefore had to leave his job and move out of this accommodation and began sharing a three bedroom house with Polish friends.

---

5   In order to be eligible for social housing, they had to provide evidence of at least one year of continuous full-time employment in the UK. This was particularly difficult for Polish builders whose employment was often casualized and temporary (Datta 2008).

Table 5.1 Benedykt's residential moves across London

Year	Duration	Type of Accommodation	Shared with	Neighbourhood
1999	3 nights	Hotel	Alone	Barking
1999	3 nights	Homeless in the Park		Hackney
1999	1 week	3 bedroom house	Strangers. Rented from 'Gypsy'	Barking
1999	3 months	5 bedroom house	Nine other strangers, Rented from Ukrainian 'boss'	Ilford
1999	3 months	6 bedroom house	Strangers	Ilford
2000	3 years	3 bedroom house	Polish friends at work	Wood Green
2004	4 months	3 bedroom flat	Rented with Irish colleagues	Marble Arch
2004	6 months	2 bedroom house	Strangers	Wood Green
2004	1 month	House	Friend	Bounds Green
2004	2 months	3 bedroom house	Strangers	Ilford
2004	2 years	House	Friends/family/ strangers	Clapton

During this time Benedykt continued to work illegally because he had overstayed on his tourist visa and could not return to Poland. In London, he was living with friends with whom he worked at building sites and who had similar immigration status as him, but very soon he began to get tired of 'seeing their faces all the time'. By then Poland was about to join the EU in 2004. Benedykt left just before the EU expansion and returned back to London as a 'legal' migrant to work in the building sector. At this point he was more mobile transnationally, and made frequent trips back to Torun, his hometown in Poland, recruiting other new migrants for a contracting company he set up with an Iranian colleague. Having already lived in East and North London for about five years, Benedykt had become familiar with those neighbourhoods and the diversity of cultures there. Thus on his return Benedykt decided to live close to these areas where he enjoyed visiting many of the Indian restaurants, take-aways, and ethnic food shops.

Benedykt's movements across different neighbourhoods of London and between Poland and England constitute both transnational and translocal geographies. These movements were simultaneously across national borders but also across different neighbourhoods, but they cannot be placed along a hierarchy of mobility. They were simply different ways of negotiating his access to particular forms of social and cultural capital (employment, language, and housing) in these places. What is significant in his mobility however is that his experiences and perceptions of London were primarily associated with particular neighbourhoods in London where he arrived in the beginning and where he slowly began to make a livelihood for himself. Benedykt mentioned that he would always prefer to live in East London because the few months that he lived in the city centre in Marble Arch was not a pleasant experience for him – it was too expensive for the quality of accommodation and too far from ethnic food shops. But his mobility across London was also because most often his accommodation was informally rented from his boss or with work colleagues; so when he moved jobs, he also had to move home. Moving between Poland and UK was similarly shaped by his immigration status – when he was 'illegal' Benedykt did not return to Torun in five years, he worked below minimum wages and was vulnerable to non-paying clients. Right after 2004, he began his own contracting company, made frequent trips to Poland and also began to rent accommodation in the private housing market.

Adrzej similarly came from Bedzin to London as a tourist in 2001. An acquaintance gave him a room in Camberwell, south London where he stayed for five months. He moved because he found another job offered by an Irish contractor with whom he shared accommodation for another year until he went back to Bedzin in 2003. After a few months he found a job in Cyprus, where he lived for 3 months and then returned back to London in 2004. When he returned he connected with a large network of Polish friends and acquaintances with whom he continued to rent cheap accommodation informally. The main reason that Adrzej gave for his frequent moves was

Work. I had to move to Lewisham [south London] 'cause I had such a deal with my partner, my business partner that, that, it was easier for me to live there 'cause we were getting to work by his car. We got on the car and went to work. And why did I move from Lewisham to here [east London]? 'Cause it was very expensive. It cost £80 a week. Let's say, I could afford it at the beginning, but then work began to be a bit problematic and I decided that [pause] I had to change it. And at the same time [friend] let me know that one room in their place was vacant. And I thought that I'd have a look. So generally speaking it's about work, or money. [Adrzej]

Table 5.2 Adrzej's transnational and translocal moves

Year	Duration	Type of Accommodation	Shared with	Neighbourhood
2001	5 months	3 bedroom flat	Shared room with Polish friend	Hoxton
2002	1 year	2 bedroom flat	Shared with English friend	Camberwell
2003	6 months	WENT BACK TO POLAND		
2003	3 months	WENT TO CYPRUS		
2004	14 months	2 bedroom flat	Friend/Polish girl	Camberwell
2006	5 months	2 bedroom flat	Irish friend	Camberwell
2006	3 months	3 bedroom house	Shared room with English strangers	Camberwell
2006	3 months	House	Rented from Polish employer	Lewisham
2006	Till date	House	Polish friends	Leytonstone

Benedykt's and Adrzej's mobilities were neither unusual nor unique. Most other participants had similar histories of transnational and translocal mobilities and most of them situated themselves in different places and at different scales through their experiences and social capital in these places. They point to a translocal notion of mobility and migration where their movements were simultaneously negotiated through the boundaries of the nation-state, the city and its neighbourhoods. London was neither an unproblematic destination nor simply the place of work. Rather the complex links between different aspects of their everyday lives produced a translocal city, where different areas of this city were continually negotiated and accessed to sustain different aspects of their life and accumulate different types of social capital.

Situating mobilities

In this particular assortment of mobilities and movements, the city was itself reworked and transformed. As they began to live and work in these neighbourhoods, participants would realize that not all neighbourhoods were affordable, indeed it was in the cheapest neighbourhoods that they would be able to afford shared accommodation. These neighbourhoods also housed large numbers of minorities – Asians, Afro-Caribbeans, and Middle-Easterns. This London that they came into was unlike what they had imagined – in the words of one participant Mikolaj, it was 'like big Babylon, you know. Every people from every country'.

In another place, I have discussed how these encounters with 'others' in London produced particular types of 'situated cosmopolitanisms' among my participants (Datta 2009a: 367), which were spatial and contextual. I noted that 'in different places, East European construction workers can be both distanciated and engaged with 'others', both cultural and political in their attitudes and behaviours towards 'others', and can be cosmopolitan, not only from the need to survive but also from an attitude towards opening up and engaging with others' (Datta 2009a). Everyday situated cosmopolitanisms of participants were incomplete efforts to understand, negotiate and rework difference within particular spaces and places of everyday life in London – drinking with English colleagues in the pubs was related to the accumulation of language skills, not meeting English women in nightclubs was related to their lack of language skills (Datta 2009b), constructing themselves as 'superior' builders in the building sites was in response to their lower position in the labour hierarchy (Datta and Brickell 2009), frequenting of Chinese and Indian takeaways was related to the relative lack of these foods in their hometowns in Poland, and the rejection of the food practices of 'others' in shared accommodation was related to unfamiliarity of smells in uncomfortably close proximity. Crucially, it was their access to social and cultural capital in these different places which determined the nature of their engagements with others like them in the city.

What emerged from my research was that different places were located within different networks of power and migrants' negotiations of difference in these places were produced from spatialized negotiations of this power. So I suggested that participants' attitudes of openness towards others in the city were neither simply a cultural aestheticization, nor just a survival strategy, but a 'complex mixture of cultural, ordinary, banal, coerced, and glocalised cosmopolitanisms that were enacted under different spatial circumstances of interaction, subjective positioning, and physical proximity' (Datta 2009a: 367). During these moments their mobilities became situated through particular ethno-national constructions of difference that were gendered, generational and spatio-temporal.

During constructions of these situated identities, the neighbourhoods where participants lived and worked underwent a reworking. London was fractured from a singular entity of the 'West' to a much more incomplete, incoherent assortment of urban sites which were spatially, socially and materially different from each other. The city centre was the place of high culture, global capitalism and modern lifestyles – places where they visited the museums and where they were employed in the refurbishment of high-end houses for the elite. The surrounding neighbourhoods on the contrary were places where minorities lived – these were also the only places where participants could afford to live. In such a reworking, London ceased to be the city of their dreams, but their local neighbourhoods became relationally constructed with reference to London as a modern city.

Figure 5.2 Dawid's picture of his neighbourhood in East London

> Dalston Kingsland is like, Hackney, Clapton, is, like for me, it's like a community centre. And people living in a really rich area like Liverpool Street, all the centre, they trying to push down these people, all these people, all these poor people. They want to make some limits, some barriers between the rich and the poor people, it feels like. [Dawid]

Such perceptions of London when compared to participants' aspirations and motivations for coming to London produced notions of a fractured city. London was both spatially and socially limiting – central London with its modernist buildings and 'rich people', and the suburbs poorly connected by public transport and inhabited by ethnic minorities. What was even more disappointing for them was that instead of living in central London and enjoying the financial successes of places like Liverpool Street, participants were living in the 'other' London, those places which they saw as deliberately marginalized by the middle-classes. What emerged subsequently was a process of identification with this 'other' London, where they lived and where like its residents, they were marginalized in their social and economic circumstances. This identification with the particular spatial qualities of East London produced a contestatory narrative of belonging and alienation evident in the desire to reside in ethnic neighbourhoods yet conscious that these neighbourhoods constituted the 'other' city.

McKay (2006a: 265) notes that 'as migration and mobility produce new subject positions, they transform and extend locality and create both new subjective experiences of place and new subjectivities'. In the case of my participants, these transformations were translocal, in that they not only made connections between their localized contexts in sending and receiving countries, they also deconstructed the notion of locality vested in the 'city'. In doing so, participants spoke of London as 'nothing special' when compared to other cities like Chicago or New York – because they had expected to see skyscrapers and modern buildings, and instead was confronted with a city which in many ways was not dissimilar from their neighbourhoods in Poland. On the one hand, their residential neighbourhoods in East London marked their disappointment and separation from this modern city. On the other hand, it was their often positive negotiations of difference in London's neighbourhoods (as in the case of Dawid) that gave them a sense of belonging in the city.

These translocal geographies produced a constant struggle in the construction of situated identities in London – 'a struggle in which discursive communities produce narratives of belonging, resistance, or escape' (Smith and Guarnizo 2006: 23). Dawid's picture of Leytonstone in East London, where he lived most recently was illustrative of such struggles. He constructed east London as part of the 'other' London, which had lost out (like him) in the financial successes of London. But he also pointed out that it is this part of London where he enjoyed living, just like he had enjoyed living in Green Lanes (in north London) when he first arrived. This was not least because he was connected to an intricate social network of friends and acquaintances from his town in Poland, but also because like other participants, he too enjoyed the ethnic food, the takeaways, and the cheap rents. Best of all, because of the increasing Polish population in this neighbourhood, there were now many Polish shops where he could get familiar food from Poland. Dawid's visual narrative thus reflects this contestation between his simultaneous belonging and alienation in the translocal city.

Situating neighbourhoods, localities and identities in the translocal city

I have tried to show in this chapter that despite heightened mobility, locality and neighbourhoods are often situated through migrants' experiences across a range of spaces. The particular feature of participants' mobility is that they are not just transnational migrants, their movements are also translocal within London's neighbourhoods, as they struggle to economize on distances between workplaces and accommodation. The building sector that they work in produces this particular feature of situatedness within London's different neighbourhoods and to the places along their migrant routes. This situatedness I have argued works as a spatialization of habitus, where access to different kinds of social and cultural capital in localized contexts produces particular constructions of places.

Thus while London is imagined as the centre of the 'West' before arrival, participants' embodied relationships with different parts of London during living and working in the city produce its fracturing into buildings, shops and neighbourhoods where interactions with others produce belonging or alienation. The fracturing of London also reflects their relationship with forms of social and cultural capital which they are able to access in a variety of urban sites. These relationships produce alienation from the modern city, even as their interactions with 'others' in the marginalized neighbourhoods produces notions of home and belonging. The fracturing of London from a homogenous entity that reflects the 'West' into a collection of different urban sites where differential access to power and capital are mobilized then produces the translocal city. Each of these places while material and embodied are also simultaneously connected and compared to other places, neighbourhoods and cities. In this chapter then, neighbourhoods and localities are anything but deterritorialized or virtual; rather their physical and material embodiment during participants' travels situates them against a range of other places and localities and neighbourhoods within and beyond urban or national space. This situatedness upholds connections between neighbourhoods in London and in their hometowns in Poland, they are both productive and marginalizing, they produce the construction of situated identities that are specific to these connections. And they allow participants to find their place within these wider notions of marginalization, alienation, attachment and affection.

Such a notion of the translocal city I suggest provides more useful understandings of migrants' negotiations of mobility and movement without subscribing to bounded/deterritorial, or national/local binaries. It allows us to examine the experiences of migration within particular localized contexts without ignoring their connections to other spaces, places and scales. And it allows us to excavate the local politics of places where migrants' 'new' lives are made through their everyday struggles to live and work like any other urban citizen. It helps us to understand their constructions of home and belonging through particular intersections between structural, social and spatial opportunities, and also to understand the city through its textures of everyday spaces and the role of these spaces in the shaping of migrant lives and futures.

Acknowledgements

I wish to thank the Suntory Toyota International Centres for Economics and Related Disciplines (STICERD) in the London School of Economics for the New Researcher Grant, which funded this research. I also wish to express my heartfelt gratitude to the participants who kindly devoted their time answering questions and taking pictures. Thanks also to Katherine Brickell and Magdalena Maculewicz for their valuable contributions towards the fieldwork.

Chapter 6

'You wouldn't know what's in there would you?' Homeliness and 'Foreign' Signs in Ashfield, Sydney

Amanda Wise

Ashfield is located about twenty-five minutes west of Sydney's CBD and forms the gateway to the city's working-class multicultural suburban heartland. It is an old federation suburb,[1] emanating a sense of genteel decay in its housing, roads and high street. Old homes with back yards dominate the southern part, while the area surrounding the Liverpool Road high street shops comprizes mostly high density apartment blocks.

Prior to the Second World War Ashfield was seen as a green escape from the urban, industrial poverty of inner-city Sydney, a place where people moved to enjoy backyards and green public space. Eventually it contained a mixture of working-class and middle-class residents, Catholic and Protestant respectively represented in the suburb's numerous churches. Up until WWII, Ashfield was almost entirely white Anglo-Celtic[2] but with the post-war migration boom, it became quite a diverse area with large numbers of Italian, Greek and Polish residents making the suburb home during the 1950s and 1960s. More recently, Indian and Chinese migrants have settled in Ashfield, the latter bringing the most profound changes to the local urban landscape.

Popularly known as 'Little Shanghai', today Ashfield is characterized by the Chinese commercial presence along its main shopping strip on Liverpool Road, home to around one-hundred small street front shops and a small shopping mall. The change has been relatively swift with the streetscape transformed since the mid 1990s from its earlier mix of Anglo, Italian and Greek shopping. Today, about eighty-five percent of the shops along the Ashfield high street are Chinese small businesses, predominantly restaurants and small supermarkets.

About a kilometre long from start to finish, a stroll along Ashfield's high street from the station towards the shopping mall is an exercise in sensory overload. Situated on a busy traffic artery leading from Sydney's CBD, trucks, buses and commuter cars thunder along Liverpool Road both day and night which makes for

---

1 A garden suburb with ornate detached homes largely built between 1895 and 1930.

2 Anglo-Celtic is the standard descriptor for the dominant majority of 'white' Australians.

a not particularly pleasant, and often bone rattlingly noisy, pedestrian experience. Popular long standing shops like the Shanghai Night restaurant and the corner newsagent are familiar landmarks in an otherwise ever changing menu of shops. To my right is Coles Photos, run by a Siberian Russian émigré since the 1960s, a little further along is 'Xin Sa' Chinese hair salon, a Chinese video and book store, the 'Go Go' Chinese supermarket, the 'New Shanghai', 'Chinese Fast Food' restaurants, the TAB betting shop, a string of $2 budget shops, a Chinese internet and gaming a parlour, and the 'Jem' Chinese fish shop. Across the road is the Chinese 'OK' supermarket, Ashfield 'Fruit World', a Chinese dumpling shop, the local pub called the 'Crocodile Inn', then the ubiquitous mall housing the large national supermarket chains. Ashfield's restaurants feature an array of Chinese food, specialties of Shanghai such as noodles, meat buns, and dumplings, enticingly displayed in windows along the strip. The Chinese supermarkets are stacked to the brim with boxes of cheap but tempting goods. Pre-packed noodles and bottles of sweet cold tea compete for space with vegetables such as Bok Choi and Gai Lan in boxes tumbling out the front doors, placed next to bulk packs of ABC tissues, cheap umbrellas and imported peanuts. It is a streetscape that lays testament to the area's new and growing mainland Chinese community,[3] who come largely from Shanghai, but increasingly from other parts of China, yet it is not unlike similar 'ethnic neighbourhoods' in multicultural cities the world over, transformed by each new wave of arrivals (Lin 1998, Mitchell 2004).

Chinese immigrant entrepreneurialism has brought a series of rapid and profound changes to the landscape of the local high-street shops since the mid 1990s. A large number of shops featuring Chinese language and shopping aesthetics have established themselves along the main street. This phenomenon has significantly reconfigured the aesthetic and material character of Ashfield's urban landscape. This change emerged as the most significant mediating factor in experiences of belonging and perceptions of difference among non-Chinese residents, particularly the Anglo-Celtic (white) elderly who were the primary focus of my study in this chapter.

---

3 I use the term 'Chinese community' cautiously, recognizing of course that communities are never stable, always porous, and are cut through with all manner of differences. Chinese in Ashfield mainly come from Shanghai (the original group who settled there), but now increasingly come from other parts of China. There are a number of Chinese dialects spoken, and also fractures around politics, and sub-national allegiances, as well as between mainlanders and Chinese from places such as Hong Kong, Taiwan, or Southeast Asia. There are also class and generational differences. However Chinese in the area tend to self refer as the 'Chinese community', thus I use it as shorthand to refer, collectively, to immigrants from China living in the area.

Figure 6.1 Typical Ashfield shops (photo by author)

New restaurants and shops selling Chinese food, music, videos and other 'ethnic' wares create a vibrant sense of home away from home for the predominantly Shanghainese immigrants in the area . These migrant place-making practices can be described as a process of creating and reproducing translocal neighbourhoods, where a sense of contiguous home is carved out which connects two or more localities – in this case Ashfield and Shanghai – both grounding transnationals and their practices within actually existing places, yet linking them across distance through material and symbolic ties (Appadurai 1996a, Smith 2001).

These translocal neighbourhoods involve intensive diasporic practices of place-making; recreating lives and spaces of embodied belonging that link to and make meaningful and relevant, one or more 'homes' elsewhere. Ashfield is not simply an ethnic enclave (a problematic term in any case). In addition to interpersonal, networked ties to 'home', the smells, sights, sounds, the very materiality of what the Chinese have created there, link Ashfield, in the most everyday ways, to the physical and sensual landscape (Wise 2010) of Shanghai. This creates, in a sense, one contiguous neighbourhood with paths – material and symbolic – tying the two places together for the Shanghainese who live there (Oakes and Schein 2006). As Smith points out in this volume, scale is also important; translocality is not always, or completely, about local-local connections. In Ashfield's case, the reference point is often Shanghai, but also includes a generalized sense of 'China', as people from other parts of China have moved into the area. Moreover, Australian non-Chinese of the cosmo-multicultural sort described by Hage (1998) impose 'Chinese-ness' as an exotic identity upon the area and this is in turn enhanced by a local government with touristic aspirations for the area.

Smith (this volume) points out that newly translocalized neighbourhoods are not empty meeting grounds. Making place translocally is a meaning-making practice... and it involves questions of power; including the power to name and claim space. Mitchell's study of Chinese immigrant entrepreneurs in Vancouver also highlights the fact that these power relations do not necessarily map straightforwardly onto the usual liberal trajectory of minority tolerance bounded by the multicultural nation state. Her study of Anglo residents of living in Vancouver neighbourhoods impacted by the place making practices of new Chinese immigrants found quite complex layers of power referencing everything from the movement of neo-liberal global capital, to competing claims to the nation state (old versions, vs. new claims to recognition and multiculturalism). There are also ongoing contestations about the aesthetics favoured by new and old residents, and these are often debated within local urban planning governance frameworks that are never as neutral as they are purported to be (Mitchell 2004).

Places are altered by these translocal practices, as I describe it later in this chapter – some more radically than others, often depending on the entrepreneurial capacities of immigrant groups residing there. Many streetscapes have been transformed by shopfronts featuring signage in the language and script of these 'immigrant entrepreneurs' (Rath and Kloosterman 2000). This is one sign of a healthy, inclusive sense of diversity, where migrant homebuilding practices (Hage 1997) of groups residing in a neighbourhood are allowed to flourish in the public domain (Iveson 2007, Sandercock 2003). Yet 'foreign language' signage is frequently a contentious issue for other residents, and often the subject of vociferous and highly emotive tabloid debate. Why is the issue of shop signage so sensitive and how might competing claims to neighbourhood aesthetics be acknowledged, negotiated and inscribed into public space?

The purpose of this chapter is to reflect upon the question of what I term 'multicultural place sharing' (Wise 2009) in the main shopping street of Ashfield and focuses on the transformation of place brought about by the proliferation of Chinese script shop signage, which represents a key aspect of translocal place making. It explores how the material, sensory and aesthetic qualities of Chinese language signage differently constitutes everyday lives and localities for the various inhabitants, producing in turn complex forms of translocal belonging and localized displacements.

The chapter is based on a broader study of 'everyday multiculturalism' (Wise and Velayutham 2009, Pieterse 2007: 98) in Ashfield, focusing primarily on Anglo-Celtic (white) senior citizens' experiences and perceptions of place change brought about by new Chinese immigrants to the area. The focus on white seniors came about for two reasons. First, as a demographic, working class white seniors are over represented among those more hostile to multiculturalism, and I wanted to understand why. And second, there is now a solid empirical literature that suggests the elderly have more difficulty in coping with place change because of a declining capacity to adapt to new circumstances and environments. This is because of declining physical health, decreased participation in the public sphere of work and family based neighbourhood life, diminished social and support networks, and for many, a psychological orientation to the past (Mair et al. 2008).

The study involved two years of intensive ethnographic fieldwork, thirty in-depth interviews with Anglo-Celtic (white) senior citizens, and a smaller number of interviews with Chinese immigrants, and other residents of Indian, Italian and Lebanese background. A number of days were also spent videotaping everyday life along Liverpool Road, focusing on the rhythms of the street, who uses it, and how people from different backgrounds interact with it as a place. The research explored a range of issues such as forms and sites of interaction across difference, multiple layers of belonging to the local suburb, the production of everyday forms of locality, of material, social and symbolic networks, and forms of neighbourly exchange across difference.

The chapter aims to understand the impact on existing neighbourhoods and residents of rapid neighbourhood change brought about by new migrants and their translocal place-making practices (Appadurai 1996a: 178). Unlike others in this collection, this chapter does not explore the translocal migrants per se. Instead, it examines Chinese language shop signage in the production of a translocalized streetscape and explores the changed landscape of Ashfield from the point of view of white Anglo-Celtic Seniors who have lived in the area for many decades, illustrating the embodied and affective feelings of connection or disjuncture this signage produces. Finally, I propose the idea of 'affective' design based interventions, informed by Thrift (2004) and Amin (2008, 2002), and argue these may contribute, if only in subtle ways, to a sense of belonging that can interweave translocal references of both the newcomers and the long term residents.

'Foreign Signs' and the Production of Translocal Neighbourhoods

Walking down Liverpool Road with Chinese residents of Ashfield casts a whole new spatial light on the strip. As I walked along with Xinnie, a Chinese community worker who was one of the earliest Shanghainese to settle in the area, she was able to tell me stories about each shop, including who owned it, when the owners came, which part of China they were from (mostly Shanghai and its surrounds), their politics and relationships with other shopkeepers and local Chinese identities (divorces, marriages, failed business partnerships), and what business interests they have in China. She also explained the names of the shops which, before then, even in transliterated form, had little meaning for me as a non-Chinese. For Xinnie, and other Chinese in the area, the names of shops along the Liverpool Road strip invoke a sense of translocal connection to landscapes, smells, sites and places in China, and particularly Shanghai. Some shops are named after famous temples, mountains or rivers in China, while a number of restaurants are named after places (such as 'Shanghai Night, 'New Shanghai', or Beijing Dumpling), or famous shops in Shanghai which evoke the quality of the food there. The Shanghai Night restaurant displays a clipped newspaper restaurant review in the window headlined 'Take me back to Old Shanghai'. Its owner, Fiona, boasted to me that her fried buns and dumplings are the best in Sydney, and that her restaurant is famous even far away in China, where a Shanghai newspaper reported about her restaurant. Another shopkeeper interviewed in the study spoke candidly of naming her shop after a famous Shanghai dumpling house 'so that people know our dumplings are as good as the best in Shanghai'. Such references are deeply translocalized as they refer to actual shops in Shanghai that many customers are likely to recognize, and in many cases would have visited. Further, Shanghai Night restaurant also features hand written signage in Russian, as it attracts Russian immigrants who lived in exile in Shanghai in the early 20th century, before fleeing to Australia post WWII. Their presence physically and in the signage of this restaurant and others, creates a sense of (complex) translocality for both Chinese and Russian diasporas who share past links with Shanghai. Shanghai was 'home' for Russian exiles, while the co-presence of these Russian and Chinese signs reminds the Shanghainese of their home city.

The majority of stores along the Liverpool Road stretch (especially clustered in the main pedestrian traffic precinct between the mall and the bus stop) have signage featuring Chinese characters. The signs are usually fairly basic in design, typically with plain red Chinese script on a white background, sometimes with a smaller English translation and/or transliteration underneath.

From a pragmatic point of view, these signs are simply a way of speaking to a target demographic. Shopkeepers interviewed for the study said that they preferred to use Chinese language signage so as to target their primary customer base. But, as I have just argued, the names also invoke places, sights and sounds of China and thus play a role in constituting the space through its connections to other places beyond the immediate locality of the neighbourhood.

At the most obvious level, the presence of signage in their own language creates a sense of place that feels manageable and navigable for Chinese residents of Ashfield, especially those with limited English skills. Chinese signage suggests to passers-by that these shops are owned by those who speak the customer's language, and ones that are likely to sell familiar wares.

> Mei: Because my mum cannot speak English...she feels very comfortable when she shops at the shopping centre because so many Chinese people there. So even though she can't speak English she can still do the shopping. ... Because there are so many Chinese people around, it is easier for her.

As Mei points out, the sheer number of Chinese shops also attracts many Chinese customers. This creates a human and physical landscape that feels familiar. Navigating a busy new place without the comfort of familiar language, shop styles or food can be highly anxiety producing for new migrants, particularly so for the elderly. Ashfield has a high proportion of older Chinese, of whom the majority are from Shanghai and its surrounds. Many of the older ones have come to Australia on a 'grandparents' visa to care for grandchildren.

> Xing Huo: In Ashfield, if you walk on the street you can hear Shanghai dialect. But not so much in 1994 but after these 10 years, lots of Chinese. They like to live in this suburb.

A familiar shopping environment, where there are shopkeepers one is able to exchange a few words with reduces their senses of isolation, anxiety and loneliness, and this in turn produces a sense of ontological security (Giddens 1991, Noble 2005: 107) in a migration situation filled with sensory, language and cultural anxiety, insecurity and feelings of displacement. Being able to hear one's own dialect 'spoken in the street' and able to see, hear, smell and consume familiar hometown delicacies actually creates a deeply embodied, sensuous feeling of belonging and familiarity, which in this case knits together both Ashfield and Shanghai.

> Lee: You know the Chinese are not used to eating sausage, sandwiches. They like to eat cabbage, soya beans. Its hard for us to change diet. So more and more Chinese came here for the food, to feel like home.

Ashfield's Chinese feel connected to the local area because it provides, in various ways, a sense of homeliness and familiarity but also a sense of continuity and connectedness to real places elsewhere. It offers a social field where their viability of being is maximized (Hage 1997). This comes from familiarity in terms of language; being surrounded by Chinese speakers, signage, food and shopping styles which all create a sense of 'homeliness' that enhances their ability to deploy themselves within their environment (Bourdieu 1990, Hage 1998, Thrift 2008) in material terms, but also in affective terms. It is all the more important because it is one of the few places in Sydney that offers such an environment.

Evoking actual places visited, or aesthetic, sensuous or imbibed landscape references, creates a deeply embedded and embodied sense of translocal neighbourhood. Locality, as Appadurai argues, has an inherently phenomenological quality, constituted by a series of links between the sense of social immediacy, the technologies of interactivity and the relativity of contexts, which is expressed through particular modes of agency, sociality and reproducibility (Appadurai 1996a: 178). Locality is materially produced through the 'organisation of paths and passages, the making and remaking of fields and gardens' (1996: 180) and the work of producing and maintaining the materiality of locality produces locality as a structure of feeling (1996a: 181) which in turn produces reliably local subjects. He stresses that locality is a property of social life (1996: 182) – i.e. figure, not ground. Even in the smallest scale locality (like a village) links and connections are drawn to elsewhere (new fields, fishing expeditions, marriages with other villages etc) which go on to expand the boundaries of the neighbourhood. As Smith points out in this volume, these links to elsewhere do not necessarily erode a sense of place. Indeed, in the Ashfield case at least, the translocal practices of local Chinese embed 'elsewhere links' into the very material fabric of the place, including through its signage. Shops that sell goods direct from China, shop names with popular references to specific places or famous shops in China, everyday business links between local Chinese and mainland Chinese business people through import-export activities, local advocacy activities through the Ashfield Chinese Chamber of Commerce targeted at the local council and local development agendas; all of these things at once create and recreate the material environment of the suburb of settlement, while embedding translocal links with specific places in China (see MP Smith 2001: 169). These connections and reference points carve out a sense of locality and place that has a particular structure of feeling and resonance for Chinese residents.

This structure of feeling is best evoked by the term I proposed earlier: the notion of migrant 'home building' rather than 'place making' practices. Hage uses the term 'homebuilding' to refer to something much wider than the domestic space, or house construction. Homebuilding he defines as 'the building of the feeling of being at home' (Hage 1997: 102), where home is an affective construct, made up of a series of what he terms 'affective building blocks'. These affective building blocks comprize a sense of security, a sense of familiarity, a sense of community, and a sense of possibility. Migrant place making practices both inside the home, and in the kind of place making practices that I have described in Ashfield's high street can be seen as materially constitutive elements of these affective building blocks. Hage emphasizes that these practices should not be seen purely as nostalgic acts of recreating the homeland (Hage 1997: 104-5). He contrasts negative with positive forms of nostalgia, the latter being the more important in creating a future oriented sense of homeliness engaged in the here and now. As he puts it, 'nostalgic feelings are sought as a mode of feeling at home where one is in the present' (1997: 104). Thus it is a mode of inhabitance that is as much materially and affectively constituent of this place and a sense of possibility and identity in the here and now, as it in linking to or evoking feelings or links to homes elsewhere. That is, nostalgic feelings and practices of this kind 'far from being an escape, are more often deployments actively fostered to confront a new place and a new time, and to try and secure oneself a homely life within them' (1997: 105).

However translocal neighbourhoods are not created on 'empty ground'. They are carved out of and involve day to day negotiations with others who occupy the place and are trying to achieve their own affective sense of 'home' in it, and in some cases involve a radical social, material and symbolic alteration of the existing neighbourhood. In the next section, I shall explore how these changes configure different forms of belonging and disconnection for 'movers' (translocal Chinese) and 'stayers' (fairly localized older Anglo-Celtic residents).

Movers, Stayers and the place making role of 'foreign signs'

Figure 6.2 The shop window referred to in the vignette below (photo by author)

During the ethnography I watched an elderly Anglo-Celtic man at the Liverpool Road bus-stop look into the window of one of the Chinese food outlets. The window had lots of unfamiliar (if you're not Chinese) small plates of food on display. The plates had tags in Chinese. And a few had rudimentary English translations – such as 'smoking fish' or 'chicken'. The Anglo man looked in the window with a somewhat confused look on his face, clearly just trying to work out what the things were. He then sits back down on the bus-stop seat. A moment later a Chinese couple come by, stand at the same window and look in. They are reading the signs on each food item, seem to be registering the prices, and have an open interested look on their face. They can obviously read the signs which tell them a bit about what the items in the window might be. If they have tried that item before then they probably already have a sensory picture of it, its feel, texture, consistency and taste, how you cook it and what you might put it with. If not, the sign will probably give them enough clues or pointers so they can approximate an idea of what it is. So already it is possible to imagine that the Chinese couple have all these ideas, labels and associations that come to mind when they view that window – associations that might entice them in, or produce in their imaginations certain possibilities about what might lie within. They enter the shop.[4] It is an encounter that mirrors how the Chinese and elderly Anglo participants typically responded on my various 'walks' along this stretch with them.

Signage, as I have suggested, is important, in practical and symbolic terms for translocal place-making and belonging. However Chinese language signage along Ashfield's main shopping street was also the single most expressed point of alienation among the Anglo-Celtic seniors in the study.

> Arthur: Its changed, when I first came here many of the shops, the fruit shops were the Greeks or the Italians. Now they've changed over recent years, they're Chinese, the big majority. Like Margaret said, you go in these shops and everything's written up in Chinese, and…you don't know what you're buying. Most of the prices are in Chinese and most of the descriptions. So you're just lost really. Why would you go in there?

I now argue that the signage issue is somewhat more complex than simply 'doing business'. Shop fronts are a major dimension of qualitative, phenomenological space, they play a key role in spatial orientation and, I want to argue, reading/ not-reading shop signage differently constitutes locality for those inhabiting and sharing such spaces.

Esther, who is 80, has lived in the same street in Ashfield her entire life. She was schooled in Ashfield, married there and raised her children there. She grew up in one house and moved only across the street when she married in 1950 to the house where she remains living to this day.

> Esther: The Chinese signs, they say to me 'don't enter'. Its saying to me 'don't come in'. 'You can't read these signs, so you're not welcome'… I mean it's a barrier isn't it.

When she expresses that for her the Chinese signs say 'don't enter', that 'they represent a barrier' what might she mean? What is the function of language but the ability to attach meanings to things? As Raymond Williams has insightfully pointed out, we learn to *see* a thing by learning to describe it (Williams 1965: 39). Belonging to a place, having a sense of it in all its dimensions requires the ability to make meaning of 'things'.

> Ruby: I don't like the fact that they splatter all the Chinese, you know, advertising or signs and they don't have anything in English. They can have Chinese but I

---

4   I spent several days videoing the everyday rhythms and encounters in and around the shopping strip during the ethnographic phase. I happened to be videoing the movements around the bus stop at the time and unexpectedly caught this moment on camera, allowing me to watch it over and over again and to observe the bodily movements and facial expressions of the individuals just described.

want English as well. We want an English explanation, so we know what it is. How do I know what that lump of stuff is?

For non-Chinese seniors like Ruby and Esther, and the man at the bus stop, the food in these windows is simply a vast indecipherable blur (or 'lumps'). Labels are not needed if they've been exposed to that 'thing' before and know what it is. But if it is unfamiliar it will remain so in the absence of legible signs or cultural translators or mediators who can explain what it is and help them get to know what these unknown items are, in an anxiety free way. Importantly, if those things are in windows then being able to read the signs mentally opens up a whole imagined landscape for that shop. Being unable to read the signs leaves that window into the shop closed. '*It's a barrier*' as Esther said.

To understand how translocal neighbourhoods are carved out of otherwise undifferentiated space, and how shop signs might play a role, Appadurai's production of locality needs further exposition. John Urry's useful discussion of Heidegger's notion of 'things' and their role in how people dwell at home and in various localities is a useful way into this issue. Urry uses Heidegger to describe how places and neighbourhoods come into being. Using the example of a jug on the table, Heidegger argues that it is not simply an instrument but actually helps to constitute the world. In this way (Urry 2000: 132) 'human existence and *[a] thing* join together in a mutual dance or play in which a world can maintain itself'. He uses the example of a bridge over a stream to explain this. The bridge doesn't simply connect the two pre-existing banks, it causes them to lie across from each other, bringing the surrounding land on either side into close juxtaposition (Urry 2000: 132). In Heidegger's view the bridge 'brings stream and bank and land into each other's neighbourhood'... it 'gathers the earth as landscape around the stream' (Heidegger 1993 quoted in Urry 2000: 132). In this way it functions to reorganize how people dwell in a place. It initiates new social patterns forming a locale through connecting different parts of a landscape.

I want to argue that, for Chinese, Ashfield's Chinese character signs, and the 'things' in shop windows themselves, represent Heidegger's bridge. By this, I mean, they operate akin to Appadurai's paths which are a constitutive element in the 'collectively traversed and legible places and spaces' (Appadurai 1996a) germane to the production of locality. They do this by drawing a sense of mental contiguity between the inside and outside of the shop, and the places and things symbolized in the signage. In turn, this creates a sense of movement into and out of the space of the shop, and the places it in turn evokes. As Urry points out, to dwell 'is always to be moving and sensing, both within and beyond' (Urry 2007: 31). As against a more sedentarist reading of Heidegger, this suggests to me materialities that interpolate bodies in a fashion not necessarily geographically bounded. Instead, it suggests that these paths or bridges, for the Chinese, particularly the Shanghainese of Ashfield, intertwine mental and sensory maps that are both proximate (into the shop) and far (to Shanghai or other parts of China). Able to read Chinese script, the Chinese in Ashfield inhabit a different sense of (trans)locality to other cultural groups in the area. It becomes a space where their sense of dwelling, the extension of their bodily self, extends not only into the nooks and crannies of those shop interiors, but into the extended translocal sensory and imaginative geographies opened up by words and associations with familiar things and places. Not only are the interiors of the shops 'opened up' and made present through clues in the signage, but the translocal connections to elsewhere are made present through shop names that reference places in China.

Chinese characters themselves also matter in the constitution of world. Often public debates about 'foreign signs' talk of *translation*, when in fact it is probably *transliteration* that matters more, or as much. Chinese, or 'foreign' script in signage makes more challenging (for the English speaker) the gradual connections that emerge when, for example, a foreign language is written in Latin script. Consider the example of the Italian salami. A number of interviewees talked favourably of the old Italian shops, and those that still exist in the nearby suburb of Haberfield, and I began to wonder why that was. I suspect much of it is a revisionist nostalgia and the accumulation over the years of a sense of gradual familiarity with Italian migrants. However conversations got me thinking about how familiarity with things Italian took place in material terms.

Fifty years ago most Anglo-Australians would have had no idea what all those 'sausages' were in the delicatessen window. Imagine, for example a deli with one of those sausages in the window, and a sign saying 'salami' in front of it. In the context of a very Anglo-Celtic 1950s Australia, where food was principally British influenced, *salami*, the word, would have held no meaning. Imagine then an elderly Anglo-Celtic woman walking past that window over weeks, months, years, seeing that sausage, and the word 'salami' spelt out beneath it. Perhaps she has sounded out that word in her head, rolled the word around her tongue and encountered it in other contexts. The point is that gradually her other senses come into play, her visual sense making the connection between word and thing. Importantly, the fact that she can read the letters means she can sound it out, connect the word and its sound to the sensual materiality of the salami. So already here, a relation is beginning to emerge.[5]

Italian, of course, is written in a script legible to English readers so there is already a process of visual recognition and sensory transfer going on when the woman tries to sound the word out, even at its most unfamiliar stages. Chinese characters, as found in Ashfield, are written in a script most non-Chinese in Ashfield are unable to read. That leap between the unknown and the known is that much greater as the Chinese script and the words represented on the shop and window remain impenetrable.[6] That produces a different relationship to the text, which,

---

5   Obviously signage is not the only factor mediating here. There are questions of relative 'whiteness', and temporal factors at play as the second generation gradually merge with the dominant majority.

6   In another context, there is a literature on 'language anxiety' in the education discipline of foreign language acquisition. Psychologists have tested the anxiety levels

in the case of the 'salami' sign, functions as the bridge, if you like, that gathers person, thing, and landscape together, despite its early foreignness. This reading suggests a debate about transliteration would be worth pursuing, where Chinese words are retained, with transliterated versions of those same words phonetically spelt out in Latin characters.

The radical erasure of Ashfield's old urban landscape – and in its place unfamiliar shops with 'unreadable' signs – removes many of those *évocateurs* of memory many elderly Anglo-Celtic seniors have laid down throughout their years inhabiting the area. There are also spatio-temporal dynamics at work, where translocal geographies of Italian and Greek migrants of the 1950s and 1960s have also disappeared, leaving curious layers of nostalgia. For example, alliances between Italian, Greek and Anglo elderly emerged in the study, and Anglos expressed a sense of nostalgic loss for the old Italian delicatessens and Greek butchers they would have once disparaged.

Generation is also important. Younger Anglo-residents express a cosmopolitan delight in the excesses and exotica of Ashfield (Hage 1997). For the Anglo elderly though, the significance of their length of residence, their very localized lives within Ashfield, and their health-induced inability to physically travel far offers some clues as to why many of them express such a sense of homely dissonance in the new shopping landscape. A health related decline in mobility has meant that movement around the city among many of the elderly residents involved in the study had shrunk from regular trips to other suburbs to weekly visits to the Ashfield high street. Those who could previously drive had given up their licences. They could not walk far and getting on and off buses, and up and down stairs at train stations or navigating large shopping malls was increasingly difficult. As a result, most of them now shopped only in Ashfield, and all their 'homely resources' needed to be drawn from the area (in the sense of neighbours, shops, and leisure). Many were socially isolated, often widows, with infrequent family visits. Shopping in the local neighbourhood was a key aspect of their social life and social milieu. Most talked of missing the 'chats' with familiar shopkeepers they enjoyed in years past. A decline in ability to grasp 'newness', and to learn about new ways of being and doing was also present. Echoing Simmel, the sheer sensory cacophony of the modern city was a source of anxiety and fatigue (Simmel 2002).

> Mary: Ashfield's just unattractive. That main street. Not one shop is attractive. They don't have pretty windows, its all busy busy busy. The shops aren't friendly. They're not welcoming. Some of the shops you wouldn't even know what they are.

> Ron: The feeling I have in Ashfield is if you have to walk down the main street, walk quickly to get away from the noise and the dirt and people's heads are

---

of new readers or speakers of a language and found elevated cortisol levels and other significant indicators of increased anxiety.

down. .... The Ashfield shops ... don't have window displays... you can't look
into the store. They have it cluttered. And the aisles are narrow and its very dark
– you just don't want to go in.

The middle class cosmopolitan pleasure of newness implies an orientation
to the future, typical among younger people. However one of the common
characteristics of ageing is an orientation to the past. Life's meaning, memories
and sense of self are drawn primarily from the past, from life lived and places
remembered. For an elderly person who has lived in the same area for seventy
years, the everyday presence of loss infused nostalgia in ones' current 'home-
place' sometimes turns to (racialized) bitterness.[7] Returning to Hage's notion of
'migrant homebuilding' as premised on a positive sense of nostalgia, he also points
out that nostalgic feelings are experientially triggered, and can be triggered by an
'experiential absence, a negative intimation, or a presence, a positive intimation'
(Hage 1997: 105). Speaking in the migrant context, he suggests a negative sense of
nostalgia is sometimes triggered by a 'direct experience of lack of homely feeling
of familiarity (a lack of practical and spatial knowledge) and lack of communality
(lack of recognition and the non-availability of help)' (Hage 1997: 106). Reading
this through the eyes of the Anglo seniors, I would like to suggest that the negative
nostalgia expressed has its roots in similar sense of loss and lack that actually has
quite a situated, practically oriented base to it. However, as Stewart has argued
nostalgia is always ideological. The 'past it seeks has never existed, except as
narrative' (Stewart 1993: 34) and thus is always remembered in the present against
a current ground. Nostalgia for the past is thus neither neutral nor factual and
calls for a return to some mythologized past need to be addressed with caution,
especially when couched in racialized terms.

Strategic Interventions for Translocal Public Spaces

Translocal neighbourhoods still occur within the national space, and in the
Australian case, multiculturalism represents a framework under which different
claims to (translocal) place can be advanced. A commitment to diversity requires
that any local intervention ensure Chinese (or any cultural group) in the area are
afforded the equal right to create and enjoy a homely environment that enhances
their capacity to 'be' (Sandercock 2003). However multiculturalism needs to be
understood not just in abstract policy terms – but as something inhabited in real,
lived, bounded neighbourhoods, shared by different groups with more or less local
or translocal orientations and senses of belonging.

I do not see the changes in Ashfield's aesthetic landscape as a mode of 'identity
expression' although obviously identities (translocal, ethnic, national and racial)

---

7   There is a literature on ageing, place change and depression in the field of social
psychology that is relevant here (Mair et al., 2008).

are evoked in the process. Instead, it represents a set of conscious practices that underpin capacities to act, be and feel at home. Nonetheless, the aesthetic quality of the streetscape, in particular, the sense of place created by Chinese language signage, does actually intimate a certain place identity of Chinese-ness. Speaking metaphorically, Amin and Thrift (2002) discuss how different spatial and temporal paths leave their marks, through the city, and these footprints (of people, paths, connections, symbols, routes) can bring 'distant sites into contact' as well as separating adjacent ones. Acknowledging that 'naming' practices – including symbolic representations and codings of the city – they suggest that people and places script each other (2002: 23). Narratives of the city are constructed through a cumulative process of 'naming' elements of it (this or that shop, park, place, street, or the local characters in it), and through this process becomes accessible, and becomes a spatial formation (Amin and Thrift 2002: 24). In Ashfield, I have suggested that this spatial formation has different intimations and resonances for the different groups who live there.

The question is how to develop an ethics of cosmopolitan place sharing that is able to encompass not only local, but translocal and transnational modes of belonging and connection. What of the other cultural groups in Ashfield, and in particular the less mobile and more marginalized among them? How to re-create a sense of connection and belonging to Ashfield among them, without taking away those things that make the suburb homely for Chinese residents and visitors? What kinds of inclusive interventions might be explored that could produce increased possibilities for positive affective connection and encounter at the interface of shop frontages?

Amin cautions that interventions at best 'come with emergent force, facilitating new spatial combinations and new rhythms of usage and regulation that will jostle against old combinations and rhythms' (Amin 2008: 22). With this advice in the background, it is nonetheless possible to imagine spaces of creative urban design, incorporating sophisticated signage and whimsical public art interventions that interweave both 'movers' and 'stayers' enabling a sense of everyday cosmopolitanism (as against a corporate or touristic version) to emerge as integral to the local identity (Young 2006). Rather than giving in to voices seeking a return to spaces of closure and locality, local authorities, inhabitants and creative workers can play a role in evolving a cosmopolitan sense of place that acknowledges and draws in both movers and stayers. These would reference their sometimes divergent belongings, visually interweaving otherwise competing narratives, practices and trajectories of place, the then and now, here and there, them and us, the local and the translocal (Vertovec and Cohen 2002: 212).

Conclusions

Translocal urbanism (Smith 2001, Smith 2005a) signals the emplacement of transnational relationships. Everyday social relations and material practices of translocal actors occur in actually existing places, reconfiguring identities, place attachments and spatial boundaries in the process (Smith 2005a: 243). Spaces once geographically distinct become intimately interconnected through the everyday fabric and materiality of place. As we have seen, these embedded translocal relations also occur in already existing places and impact upon the forms of belonging available to other inhabitants, sometimes causing tensions or competing claims of ownership. Place attachments of 'stayers' (localized long term inhabitants) are often oriented towards a more enclosed, past oriented sense of place. This is at once tied to nostalgia for mythologized pasts, but also to actual spaces and meaningful social relations that have dissipated over time. However scholars such as Massey (Massey 1997) have long entreated us not to see cultures as something mapped directly onto discrete spatial parcels. These overlapping modes of inhabitance bring particular challenges for the study of translocality. In this context, the differences between Appadurai's and Smith's approaches to translocality are less stark than at first glance. Appadurai's notion of neighbourhood does not necessarily automatically lead us to an overly deterritorialized reading. It does signal for me the importance of thinking about the embodied, corporeal and affective bonds within and between localities (Brickell and Datta, this volume) and how paths tread, and connections knitted, configure and reconfigure space, place and neighbourhood. Understanding the role of aesthetic, symbolic as well as material cultures in these forms of embodied belonging and the translocalities they configure (Tolia-Kelly 2008), adds depth to Smith's invocation to take seriously how mobile subjects are emplaced (Smith, this volume). A research agenda for translocal geographies needs to include a focus on place making and the interrelationships with both sending and receiving contexts and the multiple groups who reside at each end. Multiple scales need to be accounted for. And finally, research needs to explore how other groups residing in areas impacted by translocal place making practices may, in interesting ways, be themselves interpolated into these new material and imagined geographies and their identities, connections and orientations reconfigured in the process.

Chapter 7

Ways Out of Crisis in Buenos Aires: Translocal Landscapes and the Activation of Mobile Resources

Ryan Centner

Introduction

Massive lines of people, filling sidewalks and other open spaces, became a facet of daily life across Buenos Aires during the 2001-2002 Argentine economic crisis. 'Structural adjustment',[1] common throughout the global South, had attempted to prime the national economy for global competition through greater marketization and less state involvement, but these clusters of reform programmes approved by the IMF and World Bank were instead a disaster, yielding new political corruption, fiscal insolvency, skyrocketing unemployment and record poverty. *Porteños* – or residents of the capital city – formed queues to withdraw pesos before further devaluation, to obtain a work visa for Spain, to secure a shrinking retirement benefit, or to receive one of the new subsidies for parents in those hardest of times. Many other lines existed, and several pre-dated Argentina's catastrophic neoliberal collapse. But these examples were especially prevalent, and infused with the most palpable sense of cataclysm as well as uncertainty about the future in the once-confident, cosmopolitan capital. They were the embodiments of the pitifully third-world depths to which the high-flying renaissance project of structural adjustment had fallen. And whilst those lines generally signalled a lack of luck for Argentina, there was another queue brimming with lucky Argentines themselves.

In a time when headlines, rallies and dinner conversations alike decried the worst crisis in the republic's history,[2] I was puzzled to find a long line of dazzlingly dressed locals awaiting admission to a crowded architectural marvel vibrating with bass and Anglophone diva riffs along the old port of Buenos Aires. Situated

---

1 Structural adjustment programmes are different each case, but broadly embrace this market logic and have affected most middle-income and poor countries since the early 1980s, with many similar effects across sites (Walton and Ragin 1990, Babb 2005, Mohan et al. 2000).

2 There has been much coverage of the recent Argentine crisis. For a range of perspectives offering description and explanation, see Pastor and Wise (2001), Rock (2002), Boyer and Neffa (2004), Blustein (2005) and Carranza (2005).

on the edge of Puerto Madero's redeveloped waterfront district, this spectacular discothèque projected an undulating chiaroscuro into the humid night. Such a curious contrast of pointed glamour amidst sprawling hardship could have simply been yet another example of the stark inequality so familiar in cities, and especially in the poorer countries of the world. Yet rather than a basic case of uneven urban geographies, the discothèque and the revamped port around it were part of a very particular landscape – both physical and social, material and symbolic (Zukin 1991) – formed through booming multinational redevelopment at the height of Argentine structural adjustment from 1989 to 1999 (Centner 2007: 12-19), and permeated by a slew of social, economic and cultural resources yielding possibilities unusual for the rest of the city. The site formed a set of circumstances potentially quite at odds with the time of crisis; indeed, it was a place of latent opportunities reaching beyond the rest of the city.

The presence of fruitful social and other resources, however, does not automatically suggest their use or the realization of opportunities; instead, there are questions of practice pertinent here. Looking at that line-up of bejewelled and radiant[3] people in Puerto Madero, I wondered how they, in the aftermath of total economic crisis, were able to engage with this site to any benefit. In the pages that follow, I delineate some of the kinds of practices that allowed a select profile of Buenos Aires residents to activate mobile resources within a translocal landscape of urban redevelopment in the global South. The narrative evidence in this chapter comes from a prolonged period of field observation in the district of Puerto Madero from March 2003 to August 2005, yet represents only a small portion of a multi-neighbourhood project in Buenos Aires that entailed comparative ethnography, extensive interviewing and archival research.[4]

Both race and gender condition the opportunities available to individual *porteños* in Puerto Madero, yet this is not the central focus here because I concentrate on those who are able to use resources in the district most effectively – many of whom are men and most of whom appear phenotypically European, pointing to salient ethno-racial questions broached more directly in other writing (see Centner 2011). Like much of Latin America, racialization is prevalent in Argentina, yet it operates through a variety of registers different from Anglo-

---

3   The ethnographic impression of 'bejewelled and radiant,' refers to a set of people who fit a mainstream yet rather high-end (nearly model-like) mould of 'beautiful,' both men and women. They have the accoutrements that mark that sense of 'beauty' (e.g., the jewels and other material accessories) as well as something that goes beyond those material acquisitions, something that is much more embodied (i.e., 'radiant').

4   The interview sample in Puerto Madero totalled 15 individuals who participated in extended, semi-structured interviews according a far-ranging schedule of questions. This represents a fraction of the larger research endeavour, which also included strategic interviews with planners and politicians operating in the neighbourhood, and dozens more casual conversations that were indispensible for ethnographic insights. The area of Puerto Madero is 2.1 km² (Dirección General de Estadística y Censos 2005: 31) with an estimated population of 5,158 in 2004 (CAPM 2009).

American contexts (see Helg 1990, Garguin 2007, Courtis et al. 2009). Foremost, there is a very tight enmeshment – or even social conflation – with class, but also a lack of prominent vocabularies such as 'Indigenous Argentine', 'biracial' or 'African Argentine' (see, e.g., Warren 2009). In terms of gender, Puerto Madero's largely corporate setting – as in most parts of the world – is dominated by men in more elite positions, but there is no shortage of women in lower ranks, or amongst the more moneyed residents of the area, as a not insignificant number of inhabitants have likely inherited their wealth.[5] Whilst there are certainly gendered differences in access to and use of different forms of capital (e.g., O'Neill and Gidengil 2008), in this chapter, I follow specific men and seek to understand how their embodied experience of Puerto Madero after the Argentine financial crisis relates to translocality.

Puerto Madero as the site of research is significant to the emerging issues on cities and redevelopment during globalization, and – most centrally – the ethnographic accounts of two types of resource activation, followed by concluding remarks on the usefulness of this case and its lessons. All of these shed light on particular aspects of translocality and the usefulness of analytic perspectives that address this phenomenon. There are three main insights on translocality in this research. First, the case of Puerto Madero shows the importance of *translocal landscapes* in which resources – rather than people primarily – do the travelling. This account illustrates how Puerto Madero is part of an assemblage of sites that traverses multiple geographic scales, enabling people with the appropriate savvy and connections to benefit from resources rooted in other physical places, even if these people remain in Argentina during and after crisis. Translocality is thus not only about the movement of bodies or the strategies of states to regulate and make use of them (Smart and Lin 2007), but also how some of those bodies are endowed with the skills and sensibilities to connect productively with a landscape that is at once localized and rather globe-spanning.

Second and relatedly, this research illuminates how a particular neighbourhood – redeveloped in line with elite global norms and maintaining a range of linkages to similar sites in cities overseas – is able to experience very different fortunes than the rest of the crisis-ridden city around it. This emphasizes the fundamentally territorialized and place-based nature of translocal geographies. In contrast to Arjun Appadurai's (1996a: 33) sense of planetary 'scapes' – e.g., 'financescapes' – which represent rather particularized flows of connection, the mobile resources discussed in this chapter signify a range of flows that specifically move amongst and can become activated in certain places around the world. More precisely, they have the potential for activation through specifically emplaced practices in these sites. The conjunction of these locales through such mobilities accentuates the importance of considering them as a translocal 'landscape', which is also more

---

5   Unfortunately, public data are not available to confirm this interview-based conjecture about inheritances.

amenable to extending other concepts utilized in this chapter, such as Pierre Bourdieu's sense of 'field' (Bourdieu and Wacquant 1992), discussed below.

Third, even narrower than the neighbourhood itself, there are smaller-scaled spaces that become highly strategic in nurturing translocal landscapes. Donald McNeill (2005, 2006, 2008, 2009) has articulated the importance of specific sites and constituent architectures within the urban built environment – especially high-end hotels – for shaping the character of cities and particular neighbourhoods, whilst Ayona Datta (2009a) uncovers the microgeographies of dwellings and other physical settings than can structure the subjective experiences and material strategies of translocal practices. This chapter benefits from their contributions, but attends particularly to practical interactions with mobile resources, highlighting how people engage with translocal geographies without necessarily being migrants themselves. The next section offers further background on the unique setting for this research and the broader literature with which it grapples.

Puerto Madero: Translocal landscape amidst a context of crisis

Prior to the most recent crisis, economic reforms were changing national and very localized circumstances in quite different ways. As major structural adjustment at the national level catalysed sharp changes throughout Argentina in the 1990s (Pastor and Wise 1999), not least in Buenos Aires, this bespoke a now-familiar story: economic polarization pervading the city's occupational structure and residential areas, with social repercussions of increasing fragmentation. Several authors document these broad shifts in Buenos Aires (Mignaqui and Elguezábal 1997, Prévôt Schapira 2000, Torres 2001), and comparative research shows such trends to be common amongst the major cities of Latin America through the 1990s with similarly deepening inequalities (Portes and Roberts 2005, Grimson and Kessler 2005). In the fragmented terrain of adjustment-era Buenos Aires, there were several projects to create new kinds of globally connected landscapes, whilst other sites were left behind in the drive to become a 'global city' (Ciccolella and Mignaqui 2002, Muxí 2004).

Puerto Madero was one such redeveloped landscape incorporating global aspirations, with the remaking of the waterfront tied to early adjustment reforms of privatization (Centner 2008a: 131-150).[6] By the early 1990s, despite proximity to the political and financial core of the Capital, the port lay fenced off from the city after a half-century of disuse. The active reversal of that condition was the aim of redevelopment to closely bind Puerto Madero to the adjacent financial district, which was undergoing nothing short of a revolution during the early years

---

6 Before adjustment, there were several contemplations of how to intervene in the mothballed port, but actual use of the area was limited to scant squatters, occasional underground celebrations and secretive military operations prior to new groundbreaking in the early 1990s (Molina y Vedia 1999).

of adjustment. With the privatization of vast Argentine assets and the liberalization of transnational investment, a number of new foreign companies were establishing offices in Buenos Aires or expanding what facilities they already had. Because the existing central business district was tightly packed, and often comprized substandard facilities (Ciccolella 1999), Puerto Madero was marketed as a logical and more advanced extension of Buenos Aires's lynchpin business territory (CAPM 1999, Centner 2009: 181-186).

The result was the refurbishment of bricked storehouses along a string of wharves to suit high-end office, restaurant and residence purposes. Across the locks, new gleaming new constructions filled former railyards and other stretches of portside land, mimicking the style and scale of the storehouses but in steel and glass façades. French, British, American, Dutch and Mexican corporate clients – amongst others – fill these buildings today as during the 2001-2002 crisis.[7]

This redeveloped landscape is also a translocal landscape because it remains connected to circuits and flows that link it with other cities around the world, from Barcelona to Sydney to Miami. An emphasis on translocality – as Michael Peter Smith (2001: 169-171) contrasts to transnationality – is crucial because it highlights exactly how Puerto Madero is a very localized milieu that remains tied to other locales more so than the rest of the nation in which it is situated. In conjunction with Sharon Zukin's (1991: 16) sense of landscape as 'an ensemble of material and social practices and their symbolic representation', we can also understand the neighbourhood of Puerto Madero itself as extending, in practice, beyond its physical perimeters on the *porteño* waterfront to encompass the far-ranging flows amongst distant locales to which it is connected. As rightly noted by Tim Cresswell (2003) and others, it is important to keep an analytic focus trained on practice within landscapes so as to avoid fetishization of their projected image; this research sustains that imperative by focusing not simply on the flashy aura cast by the redevelopment of Puerto Madero, but instead on how some *porteños* are able to make the fleeting translocal flows of the neighbourhood work for them in practical terms.

Although the specific location of this account within a translocal and unusually privileged landscape amidst an otherwise downtrodden Buenos Aires is fundamental to shaping the story of crisis and how some manage to weather it, there is a more general relevance that should be underlined: in an era of widely circulating ideas, people and resources, and with the reality of severe global-economic turbulence, it is especially helpful to understand why some groups in some places are able to emerge better off, to find a way out when others cannot. By concentrating on Puerto Madero in its unique positioning during crisis, the chapter highlights what Jennifer Robinson (2002: 545-546) calls the 'ordinary' aspects of city life but for a very extraordinary neighbourhood and set of beneficiaries; which

---

7 Since roughly 2005, there is also a growing backdrop of luxury apartment towers and an emporium of premium hotels emerging behind the new constructions for corporate clients. However, these had not yet materialized during the crisis.

indicate how the endeavours to redevelop Buenos Aires as a global city have been so precarious yet consequential. At the same time, I pursue Ananya Roy's (2009: 820) cue about using this kind of irrevocably peripheral account to theorize the nature of translocal urban landscapes. Rather than a mere tale of privilege, Puerto Madero shows the significance of where and how fast-footed resources become useful. Several broad concepts, elaborated in the following section, make possible the illumination of these lessons in both Buenos Aires and sites farther afield.

Emplaced practices and mobile resources in a translocal landscape

As already noted, several researchers highlight the importance of understanding landscapes in terms of the practices that contribute to composing them. To better apprehend such practical aspects in the case of this redeveloped neighbourhood pervaded by potentially beneficial but fleetingly mobile resources, it is helpful to enlist another framework that grapples with the practices of using and acquiring different forms of capital within particular fields of action: in this instance, within a translocal landscape.

Bourdieu's set of flexible concepts fits this task especially well. According to his framework, practices are simply the acts carried out by people with different capabilities, dispositions and goals, but are seldom entirely rational or planned because 'the logic of practice is logical up to the point where to be logical would cease being practical' (Bourdieu, as cited in Wacquant 1992: 22-23). Therefore, people seek to accumulate and expend resources based on their engrained dispositions that animate practices (Bourdieu 1990: 56), but these practices occur within the extremely fuzzy logics of affect (Bourdieu 1990: 86), and amongst rather inegalitarian constellations or fields of possibilities (Bourdieu and Wacquant 1992: 94-97). Such dispositions draw from a lifelong inventory of experiences and advantages or disadvantages,[8] which then guides judgements of what is sensible action in a particular field.[9] In a translocal landscape, the field extends to a range of locales across scales; this case presents people operating with translocal strategies and practices even as they themselves are largely immobile in the Buenos Aires crisis context.

The mystery that opened this chapter – about how people could possibly benefit from lustrous Puerto Madero in the middle of Argentina's worst economic crisis

---

8   Bourdieu refers to this guiding inventory as 'habitus' (see also Wacquant 2005).

9   This set of concepts allows for great portability across quite distinct sites of analysis, and certainly pertains to questions of geography and place (Painter 2000), but Bourdieu is particularly resistant to spatializing his own framework (Centner 2008b: 197). Furthermore, despite his strong interest in addressing movements and dilemmas related to globalization (e.g., Bourdieu 1999, Bourdieu 2000), Bourdieu's application of his notion of field generally falls short of embracing any scope beyond the nation (e.g., Bourdieu 2001: Postscriptum).

– can thus be solved by looking at the practices deployed by those who manage to use this translocal landscape to their own gain. It is not always a matter of the exact physical installations in the redeveloped neighbourhood, but these do play a key part in the stories portrayed here. Rather, the built environment is part of this field of opportunities, which channels a number of mobile resources through them by attracting particular people, their different forms of capital, and the possibilities that attach to them. Most crucial here, analytically, are (1) the dispositions and already-existing stores of advantages that enable certain people to activate these resources that so many in crisis-ridden Argentina find painfully out of reach, and (2) the specific emplacement of practices in this translocal landscape that make such activation fruitful.

Two accounts from Puerto Madero yield empirical insight onto these issues. The people, practices and locations in each differ, rendering distinct angles on the kinds of ways out of crisis that some were able to procure. These portraits derive from research fieldnotes to relay, first, how *porteños* piece together resources translocally in an elite hotel and, second, how they metaphorically slide between locales within the geography-bending experience of that fortuitous nightclub.[10] These illustrative examples demonstrate some of the key emplaced practices for activating highly mobile resources in a translocal landscape rocked by deep crisis.

Piecing Together Resources Translocally

The first time I encountered Darío Calabrese, 31, was in March 2003 at the very outset of my fieldwork. Meeting him, however, had not been my goal when I set out for the hotel where he was working at the time. A friend of a friend of a friend from San Francisco named Mitch – staying at the Hilton Puerto Madero after several previous visits – wanted to give a boost to my research by introducing me to his local acquaintances at the hotel. So I went to meet Mitch and see where his contacts might lead. Tucked between two passageways around the corner from the newly relocated Dutch embassy, the Hilton had become a cultural icon in Argentina in the handful of years since its construction. Before I even stepped inside, I already knew the interior from numerous films that used it to portray lavish luxury (see Jusid 2002) as well as technological sophistication (see Bielinsky 2000) in Buenos Aires.

My arrival as a pedestrian, rather than by taxi, obviously surprised the staff; probably my very basic clothes and mid-20s age also contributed to an out-of-place ensemble appearance for those greeting new arrivals. Before the sliding doors released a blast of air conditioning into the palpable *porteño* humidity of autumn, I was met by inquisitive hotel workers whose faces suggested, 'Who the hell are you?' Accustomed as I was to greater anonymity in hotel lobbies, I went straight to the reception, where agents demanded my passport, examining it carefully. Whilst

---

10   To protect the privacy guaranteed to participants in the course of field research, all personal names have been changed in this chapter.

waiting, I noticed a long series of security monitors on the back counter, with an apprentice learning how to watch the screens with careful exactitude. In perfect English, the Hilton staff notified me that Mitch was in the VIP lounge upstairs. They called over Darío, who recognized Mitch's name immediately, but referred to him strictly as Mr. Doogers. After an overly courteous greeting, he chaperoned me through the Hilton to be Mitch's guest for cocktails.

Up the sleek glass elevator, the entire hotel atrium was in sight, as were the port and financial district's skyscrapers beyond the clear encasement of the building. Taking in the view, I tried to make smalltalk with Darío in Spanish, but he clearly preferred limiting our conversation, and keeping it in English. Still there was a stiltedness, as if he were on guard. At the top floor, he showed me through to a wide private room with several liberally spaced tables, some couches, a widescreen TV and a long bar of all imaginable high-end spirits in easy reach. Aside from two Mexican girls talking amongst themselves, Mitch was alone at a table speaking with Mariano, another Hilton employee, who was pouring Mitch a whiskey as we approached. After some pleasantries in English, we were soon joined in the increasingly friendly banter by another guest, Mandeep, an economist from the World Bank in town from Washington, DC, to work for a few weeks with the regional office at the doorstep of Puerto Madero.

I learned from Mitch that this was his eighth visit to Buenos Aires within a year, and that he had befriended many of the staff at the hotel. Only one of his trips to Argentina was actually for business; the rest were for pleasure – for the wonderful people and the cheap peso, he told me. Although he had met several expatriates in the Hilton, especially European executives and American flight attendants, Mitch was more eager to test his Spanish with the staff, who seemed an equally eager audience for his many stories. Listening to Mitch's stammering struggles with Spanish required much patience, yet its potential payoff was significant.

Patience, indeed, proved quite profitable for Darío and Mariano in particular. Mariano was in his late 30s, from a lower-middle-class family in the western part of the Capital, working full-time as a bartender in the hotel. The pay was mediocre, but Mariano explained that the perks were in the environment and the company of both his co-workers and the guests. Yet, he was about to leave his job. Mariano had recently acquired an Italian passport through his immigrant grandmother's birthright and the counsel of another guest he came to know well. Mitch, however, was helping Mariano with a free frequent-flyer ticket to the United States – where he could now travel without a visa due to his new European documents – in his search for a more lucrative job before going on, again courtesy of Mitch, to Spain and Italy for the same purposes. But it eventually became clear through further interactions that Darío was gaining even more through his presence at the hotel.

Darío came from the working-class suburbs south of the Capital. He was always a high-achiever in school, and managed to begin studies in medicine within Argentina's free and high-quality but enormously bureaucratic public university system. Even whilst staying with his parents to save money, however, it was increasingly infeasible for Darío to attend university at night and get ahead in his career. In 2000, he found out from a friend studying tourism about openings at the Hilton, where even his small amount of academic English was in great demand. He was excited to take the better-paying job, but its long and irregular hours definitively required he withdraw from university. Studying to be a doctor had become an apparently impossible dream anyway, Darío figured, especially with the downward direction of the Argentine economy.

But Mitch had a particular liking for Darío, as an especially pleasant friend but also one in need of help. The guy I had encountered earlier in the elevator, so shy of smalltalk, was like a different person in his interactions with the clientele, but especially the several times I saw him around Mitch. Married, childless and middle-aged, Mitch took to Darío like a surrogate son, and found his tale of hardship fascinating. He decided to funnel some of his surplus earnings from the Silicon Valley to help Darío leave his job, rent an apartment in Las Cañitas (an expensive neighbourhood by *porteño* standards, but very cheap compared to Bay Area rentals at the exchange rate of 2003-2004) and attend the private, more pragmatic Universidad de Belgrano. In fact, Mitch's last trip was meant to make sure Darío was established and ready to resign from the Hilton in time for the new academic year.

Of course not all employees at the Hilton were anywhere near as fortunate, nor all guests as generous, but the hotel represents a site in Puerto Madero where it becomes possible for those with just enough of an accumulated resource — be it education, a European grandparent, or a knack for customer service[11] – to engage with and activate mobile resources that loop through the translocal landscape of Puerto Madero, but might otherwise simply pass them by. In Darío's case, he was able to radically change his own trajectory by piecing together resources from altruistic travellers as well as local acquaintances with useful advice. He skilfully apprehended all the cues in interacting with foreigners and people with far greater material resources than his own. Whilst a somewhat extreme case, it is important to emphasize that Darío is neither particularly manipulative nor unusual. His location in Puerto Madero, where he could fruitfully activate mobile resources transiting amongst more fortunate locales, put him in a very opportune position.

Darío's practices, therefore, assemble and activate the fast-moving resources of Puerto Madero without sacrificing his own aims. He did not fit too well upon arrival at the Hilton, but through improving English, a connection in hiring and helpful co-workers, Darío was able to engage with the new environment just enough to seek out his dreams – even during the worst of the economic crisis that devastated the Argentine adjustment model but could not undo the glamorous redevelopment of Puerto Madero. In this mode of learning new dispositions, Darío also took the many compressed lessons of the Hilton with him to the Universidad de Belgrano where he found himself already fluent in the languages – foreign or

---

11   Although these are ethnographically specific examples, conceptually they represent various states of Bourdieu's forms of capital and habitus.

otherwise – of the more privileged, and poised to maintain useful future ties with the hotel and its staff.

Darío's account points to some key lessons about translocality regarding both the built environment as a strategic configuration and how the crosswise connection of locales creates circuits of mobile resources that are much more specific in their flow than a global span or nation-to-nation trajectory. McNeill (2008: 390-392) situates prominent hotels within cityscapes as important anchors of development and redevelopment as well as facilitators of various forms of circulation – of bodies, capital and ideas – to and from other places. This case shows how the Hilton Puerto Madero is indeed an iconic place, but that its positioning and particular milieu fosters the circulation of potentially fruitful resources well beyond the city itself. The hotel therefore becomes part of multiple far-ranging translocal scapes (Appadurai 1996a) but it remains fundamentally rooted to the redeveloped waterfront district as a rather circumscribed space of opportunity. Whilst this kind of experience in the hotels of Puerto Madero signals the substantial transformation of selves through interaction with the translocal landscape, the following section shows how others were using the neighbourhood to reconnect with, and hopefully regain, resources from other parts of the world.

Sliding Between Locales Productively

Cristóbal García-Riveux, 23, greeted me with a nervous smile. 'All these people out on a Wednesday night – this is unbelievable,' he thought aloud as we joined a long queue snaking around Opera Bay. Although we had already met several times through mutual friends in a range of social settings in Buenos Aires, he seemed just a touch uncomfortable in the dressier clothes he wore to work simply so he could come out for drinks in Puerto Madero at the end of the day – except that at 7:00pm, the typical *porteño* workday still had an hour to go. 'This is so early, nobody finishes work by now!,' Cristóbal exclaimed, and then his face glazed over with shock as he realized a good number of those 'beautiful' people in their 20s and 30s, lined up for *el after-office*, had not come from any offices at all.

To fit the advertized dress code of '*presencia excelente*' for each attendee, in an '*elegante-sport*' style, men and women had put on their suits – many of them coming from their homes – to take part in this boisterous but highbrow happy hour. As we inched toward the soaring, sleek fractal-like reinterpretation of the Sydney Opera House (see Figure 7.1), with its music throbbing and lights casting dreamlike hues of dark purple-pink, Cristóbal found it hard to fathom how the rest of these apparently jobless revellers could afford their way. After all, it was early 2004 and Argentina's famous crisis was still a very stark, haunting reality.

Figure 7.1 madero½week at Opera Bay (photograph by author)

Even with his job as a bilingual computer consultant at a nearby call centre for an American software firm, he was about to spend half a day's wages just to enter and receive a ticket for two mini-bottles of champagne. Ordering *hors d'oeuvres* from the exaggeratedly polyglot food menu would have been beyond reach, let alone pronunciation. But it was an event Cristóbal had been waiting for, a place he had long wanted to experience, so without thinking more about it, we paid our 20 pesos each and stepped into a new world.

Inside, '*madero½week*' – which is not a translation – was a spectacle of the Buenos Aires would-be '*yupi*' [yuppy] at play. Throngs of lithe women and strapping men surrounded tables and inconveniently placed but dazzling pools of water throughout the four rooms, two levels and three outdoor patios that looked onto Puerto Madero's marina and the financial district known as *La City* (see Figure 7.2). Upstairs, Cristóbal and I eventually found some room to hear each other over the British and North American soundtrack of electronica and 1980s pop. We kept an eye out for our late-arriving friends from a large Chilean corporation and the Australian and German embassies, but in that blend of bodies below, it was hard to tell anyone apart.

Figure 7.2 La City, seen from Opera Bay (with World Bank in the crown of largest visible skyscraper) (photograph by author)

As the revellers themselves would say, '*el look*' of *yupi* similitude was no small achievement given the diversity of circumstances constituting the crowd. It turned out the mystery of the celebratory unemployed was, in fact, more complicated than had appeared outside. Amongst the attendees there were those with skilled middle-income jobs like Cristóbal, who had to pull strings to leave the office early in order to arrive before the cover charge doubled. But there were also many attendees making even bigger proportionate investments from their erratic incomes just to partake in *madero½week*, as well as others who were not working at all because they had easy access to sufficient family funds bankrolling their posh lifestyles.

Amongst the crowd, there was a late-20s accountant from Caballito, his devil-may-care playboy buddy from San Isidro, and their unemployed university classmate from Belgrano. They all stood in a clutch with loosened ties, wavy light-brown hair and broad shoulders shaped by rugby, sipping champagne whilst they shared impossible stories about '*minas divinas*' [divine young women] and kept watch for '*esa nena re fachera*' [that really eye-catching girl] in the corner.

Opera Bay is where those friends come to catch up with each other, but also to look for new girlfriends and new jobs. They were, for the most part, at ease there – in those clothes, with that music, amongst those women, talking politics and business in between jokes. It was an environment that suited their expectations as someplace familiar, even though there was nowhere else in Argentina quite like it.

Cristóbal's experiences were somewhere in between this spectrum of friends, all approximating the same image. After our initial visit together, he attended Opera Bay with some frequency for several more months but eventually tired of its bustle – the event, however, far from dissipated, only growing with time. During his period of frequenting *el after-office*, Cristóbal was not one of the unemployed, but easily could have been due to his patchy and short employment history as a young worker, if the American software company had not greatly expanded its global call centre adjacent to Puerto Madero after the peso devaluation. He therefore had less to gain, in a sense, than the jobless in suits. However, he did make contacts at Opera Bay that helped lead to some of his later employment pursuits in two of the most elite hotels in all of Buenos Aires.

This middle-road trajectory relative to other partygoers did not change the fact that they all shared a similar set of familiarities, with Cristóbal and the less fortunate attempting to better align with the Opera Bay ideal. It is important to underline that this ability to approximate is grounded in substantial previous exposure to, or sometimes even achievement of, that cosmopolitan point of reference. As an example, Cristóbal, despite being only in his 20s, had an extensive earlier connection to the base knowledge of foreign popular culture and English that gave those at Opera Bay a similar foundation. Certainly it helped that in the 1990s, with deregulation and privatization, an abundance of overseas programming became easily available to anyone who could afford cable television. But an even more exclusive and intense engagement came during a year of high school Cristóbal spent in Long Island, New York, when he became completely bilingual, with a mammoth repertoire of adolescent slang, and gained an affection for such oddities as Swedish fish candies, Gap sweatshirts and Manhattan trinkets. None of this would have been possible without growing up in a part of greater Buenos Aires like San Isidro, attending a high-quality and internationally oriented high school and having a number of globetrotting acquaintances.

A great number of the *porteños* at Opera Bay hailed from similar upbringings and specifically 1990s experiences of travels abroad with a strong peso that enabled young, upper-middle-class Argentines to become the first generation – outside of elites and dictatorship exiles – to see the world. High on the list of temporary foreign destinations for mobile and fairly moneyed *porteños* before the crisis were Miami and London (Melamed 2002), where they became acquainted with entirely different worlds of music, fashion and language. Whilst somewhat superfluous, such acquisitions became part of these young Argentines' ongoing store of experiences, orienting their tastes and actions upon their return to Buenos Aires, and suffusing imaginations of better times and better places that would inform their ensemble of coping strategies in the worst of crisis. At Opera Bay, they could return to the embrace of those preferable locales without ever leaving Buenos Aires. More usefully, they could bond with others who shared similar experiences abroad, and activate their privileged past as a resource for gaining new personal connections to find improved work or any work at all. These are manifestations of strategies that are fundamentally translocal, reflecting a habitus that is familiar with a range of sites and practices far removed from Buenos Aires, yet useful in this kind of particularized environment in the Argentine capital.

The trajectory for many young revellers had been one of great promise followed by deep disappointment that places like Opera Bay could thus potentially salvage. There were ever-increasing expectations during adjustment, then dashed by protracted recession and cataclysmic devaluation through which many of them and their families lost their savings – at least in the sense that had come to matter to them: the value of those savings relative to foreign currency markets. From this perspective, Opera Bay and other sites of leisure in Puerto Madero were venues of foreign-infused effervescence that many sought to recapture and enjoy, but also presented strategic sites where having the savvy to slide between different worlds could be sufficient to enter or re-enter more privileged ranks through important material gains facilitated by the connections forged there.

Conclusions

Whether assembling an array of resources translocally or endeavouring to maintain a connection to experiences and repertoires that straddle locales, these two accounts portray some key ways out of the Argentine crisis. The first case, illuminating Darío's – and others' – practices of piecing together opportunities from a range of resources moving in and out of their purview, shows how an elite hotel created a networked translocality for workers to activate various forms of capital from locales abroad (e.g., San Francisco) that are highly valued in Buenos Aires, enabling safeguards and even improvements during Argentine crisis.

The second case, revolving primarily around Cristóbal's nocturnal ventures in Puerto Madero, shows how a particular discothèque helped maintain elite linkages to overseas locales for the purposes of both basic nostalgia and to navigate personal ways out of crisis. Although merely an entertainment site on the surface, this account demonstrates its status as a crucial translocal venue capable of keeping aspects of more promising elsewheres (e.g., representations of and resources from Miami and London) attainable even during dismal times in Buenos Aires.

'Times of crisis', as Achille Mbembe and Janet Roitman (1995) argue, yield enormous if uncomfortable or heartbreaking contexts of creativity. This encompasses resourcefulness, a characteristic to which all can resort, yet in a vast city like Buenos Aires, the resources on which people can draw – from the past and in the present – are far from identical across the variegated urban population during the upheavals of crisis. In both of these accounts, it is those with rather significant existing stores of resources before crisis who are able to manoeuvre this translocal landscape most adeptly and activate mobile resources. Their efforts to scramble and accommodate to the brusque change in the field of opportunities and constraints is akin to the 'dancing' that Aiwha Ong (2007: 7) describes amongst adaptive elites in the radical makeovers of Southeast Asian and Chinese urban economies, a term she uses to refer to shrewd practices of moving artfully across quite different spheres and logics of value. In much the same way, these *porteños* are dancing productively across the translocal landscape of Puerto Madero; again, this is not only the physical neighbourhood itself, but also the field of overseas locales through which numerous precious resources circulate during times of crisis.

As part of a translocal field of resources, this neighbourhood and its multiple privileged connections abroad thus constitute a multiscalar landscape that can be activated by certain urban social groups with the know-how for directing their emplaced practices beyond Buenos Aires, whilst it remains inert and essentially resourceless for others who cannot effectively exploit translocal connections. The locale of Puerto Madero is, in effect, an example of a 'space of exception', but not at all one of extreme disadvantage or suspended protections (e.g., Al-Sayyad and Roy 2006: 14-15). Instead, Puerto Madero was a site of unusual opportunity – for those endowed with the savvy of previously accumulated resources – relative to the rest of Buenos Aires during crisis (see Sarlo 2008). Certainly there were other means for grappling with the hardships of crisis that were not primarily translocal in their scope, but these typically did not entail quasi-permanent solutions or triumphant improvements (e.g., Auyero 2007, Whitson 2007).

An important working-class contrast that is somewhat translocal in its appeal, and especially relevant to the hotel case presented here, can be found in the Hotel Bauen collective in central Buenos Aires (see Evans 2007, Faulk 2008). As part of a broader movement of worker-recuperated businesses in the wake of the 2001-2002 crisis (see Lavaca 2004, Rebón 2005, Fields 2008, Vieta 2010), former employees of the Hotel Bauen took over the bankrupt, shuttered enterprize as collective owner-workers. This was a venture that relied heavily on a solidarity network of similarly minded movements based in an array of sites – mostly in Argentina but also abroad – for survival in the face of growing legal and political opposition (Faulk 2008). In this way, the Hotel Bauen provided even broader-based if less illustrious opportunities for *porteños* to persevere through the worst of crisis than the exceptional sites in Puerto Madero.

Other innovative practices emergent amongst the poorest *porteños* during crisis were also translocal in their pitch, such as the wide-scale rise in *cartoneros* or scavengers who recycled rubbish throughout the city and sold it to factories that, in turn, paid higher rates than ever before due to the global (i.e., not *peso-*denominated) market for recycled paper and metal (see Anguita 2003, Chronopoulos 2006). Although not a means for generating great wealth, this translocally minded strategy has provided subsistence for tens of thousands of Argentines since the onset of crisis and during the tumultuous recovery years. Relative to these kinds of initiatives, an elite, redeveloped waterfront district is quite exclusive. However, by analytically prying inside Puerto Madero, it is possible to scrutinize a much less studied microgeography of translocality, and one with characteristics that are often glossed as successful in some unqualified, inevitable way, due to the privileged and global orientation of redevelopment in such sites.

Indeed, most *porteños* are locked out of accessing Puerto Madero in the most mundane ways (e.g., González Bombal 2002: 108), let alone activating its resources. Yet even the more privileged in these accounts have to 'dance' in order to make use of the translocal landscape and its opportunities in a period of total calamity wrought by structural adjustment. It is not necessarily they who travel, but their resourcefulness must be mobile enough to make connections across a range of helpful sites and flows from them. This is especially relevant for cities in the abundant cases of developing and middle-income countries that have entered deep crises after failed projects of structural adjustment reforms. These cities, and in particular such circumscribed areas of globalized redevelopment within them, fostered myriad useful connections with strategic overseas sites during the height of reform, which can still be useful to some residents even after global-economic links atrophy, but only through the deft manoeuvres of practices that successfully activate translocally mobile resources.

Residents of Buenos Aires are seeking a way out of crisis without actually leaving the crisis context – at least not yet. They are attempting to finesse some utility out of resources from elsewhere as they become most readily accessible in particular venues of the translocally connected district. But these complicated efforts underscore how translocality must be practised in order to be useful; it does not simply deliver utility. Translocal geographies of Puerto Madero residents, are not only defined by the connectedness of far-flung, specific locales, but the everyday and ongoing practices that suture them together and draw on resources circulating amongst them. In these connections of practice, there are the possibilities for creating ways out of crisis but they are highly specific in terms of where these resources-as-possibilities become realizable and who has the accumulated dispositions, or habitus, to activate them effectively.

Acknowledgements

I thank Brenda Vacatello, Pha Pha Hamilton, Cristián Daghelinckx, Federico Hahn, Juan Robledo, Randall Stieghorst, Kirsty McNeil, Ian Duddy and Tanja Roy for facilitating fieldwork in Puerto Madero, and Michael Austin Sui for indispensible help with perfecting the document images. I am also grateful to Ananya Roy, Mark Healey, Laura Enríquez, Carmen Rojas, Ryan Devlin, Hiba Bou Akar, Jean-Paul Vélez, Pedro Peterson and the late Stephanie Kim for their crucial input on earlier drafts of this writing. Research for this chapter received financial support from the University of California Regents and the Andrew W. Mellon Foundation Fellowship in Latin American Sociology.

PART 4:
Urban Translocalities:
Spaces, Places, Connections

Chapter 8

Fear of Small Distances: Home Associations in Douala, Dar es Salaam and London

Ben Page

Who's afraid of the local?

In his 2006 book, *Fear of Small Numbers*, Arjun Appadurai analyzed the aspects of globalization that have led to violent forms of ethno-national identity. The 'small numbers' to which he referred were minorities within nation-states, groups of 'migrants' who are the outward sign of the invisible processes of integration at ever greater scales. These groups generate fear, he suggested, because they can reveal that any attempt by a majority group to claim that their own identity is *the* national identity is generally a lie. In this account fear is the precursor to violence.

In contrast the literature on translocalism (also pioneered by Appadurai 1995) tends to take a more positive position on the emerging forms of social relationship that are associated with contemporary forms of global integration (Peleikis 2000, Smith 2003, Velayutham and Wise 2005). Fundamentally it does so by evading the register of the 'national.' In part this reflects the simple (but important) empirical observation that for many people national identity is of limited importance (Schueth and O'Loughlin 2008). However, the deliberate intellectual goal of pushing questions of nationalism into the background is based on another premise too. The claim is that bonds and bridges formed by translocal relationships may escape the mire of trans*national* politics in which rigid and antagonistic identities are defined in terms of ethno-national belonging and where the engagements of the territorially distant can fund acts of violence or prop up questionable political regimes 'at home' (the role of the Eritrean diaspora fuelling the war against Ethiopia is a useful example (Bernal 2004)).

According to this perspective, 'place' is not bounded, unchanging and exclusive, but is amorphous, dynamic and open and this is particularly true of urban places. The claim of translocalism is that mobile people, money and ideas can generate a new way of developing a place-based identity that makes limited reference to the nation. Such identities draw on more than one site but retain earlier readings of place as 'space plus meaning'. A place-based identity can be characterized as 'translocal' if an individual's notion of their home is constructed out of more than one 'locale'. This personal project of translocal home-making, it is claimed, is becoming more common.

By definition the different sites that are imagined in a *trans*local relationship are assumed to be physically distant, so the title of this chapter appears paradoxical. Where are the 'small distances' to which it refers? One of the possible circumstances of translocalism is of spatially widely dispersed groups of people who are involved both with the specific place where they live and also with a distant place (such as a natal hometown) that they all have in common. Their social relations in the place where they live exist in a separate register from the social relations that are organized around their hometown. In relation to that hometown these people act as if they were geographically proximate even though they are geographically separate. In this sense translocalism is synonymous with 'non-geographic localism.' The 'small distances' come from the idea of behaving *as if* distances were small, even though they are not in geographical terms.

But why should such a focus on the small distances of the translocal generate fear? A focus on the local is a cause of worry because local politics is often rather reactionary, not least because it undermines attempts to establish solidarity at a larger scale on other axes. In its enthusiasm to dispense with the problems of nationalism, does translocalism gloss over the dangers of localism?

Localism is an ideology. It establishes a normative view of the merits of the local that often turn out to be in the interest of particular powerful social groups. Localism associates the local (often treated as the village, small town or suburb) with a set of descriptors that are taken to be worthy of defending: mutual support within a coherent community, high levels of social interaction between small numbers of people, transparent lines of accountability because of the small scale of politics, a perpetuation of tradition, a mechanism for social continuity and a society marked by high levels of shared knowledge and values. Sometimes it is not only a description of, but also a defence of, political and social conservatism. As such it tends to endorse the status quo and is reticent about the ambivalent consequences of lauding the rights/duties of the community members. There is value in thinking about 'localism' as an ideology not only because it raises important (if familiar) questions (such as 'in whose interest is it to perpetuate the status quo?') but also because it draws attention to the delusion that small scales have causative powers: there is often a spatial determinism lurking under claims about localism. So, despite its association with global scale movements, translocalism might be linked to the fears of small distances because of its commitments to the local. In an African context, for example, local politics is particularly associated with the dangers of reinforcing ethnic segmentation and of reasserting a politics of culture and belonging, which is based as much on exclusions as inclusions. Yet this gloomy critique of the local always imagines a community that is socially uniform and isolated in character.

Conversely, this chapter uses case studies from Cameroon and Tanzania to argue that not only are individuals often meaningfully committed to multiple urban communities, but a particular place usually turns out to contain different social groups and is not socially uniform at all. Insofar as it helps to reveal such stories, translocalism is a valuable analytical device. Furthermore whilst the narrative that emerges is one in which migrants' expressed values often reflect a commitment to the ideology of 'localism' (even when they themselves are widely dispersed) it is premature to dismiss 'translocalism' as necessarily reactionary for this reason. Rather (like transnational politics) there is always the potential for the practices of translocalism to be signified by 'openness' rather than 'exclusion' (Massey 1993, 1994). Second, it argues that an *historical* analysis of migration and urban loyalties reveals a long-term story in which the place-making activities of the past have often drawn on multiple locales. The relative importance of different local homeplaces changes over time, but for many people it was never a zero-sum game in which one urban place erases another in the affections of a family or individual. This part of the argument raises a different question about translocalism by questioning its novelty.

The next section of the chapter describes a journey between London in the UK and Bali Nyonga in Cameroon in order to show how an individual's place-making projects draw on multiple locales. It is an amalgam created from many of the stories we were told in fieldwork interviews rather than a specific journey made by a specific person. The section illustrates the inappropriateness of thinking about distinct cities as opposed to an urban process and it highlights the different emotional and material connections that emerge from the project of pursuing local concerns over long distances. The third section of the chapter gives a brief description of the work used to research the case studies before using the empirical material to answer the questions: have cities become increasingly significant as sites of affiliation for international migrants relative to nation-states? And, why is it that cities are so key to producing and reproducing translocal linkages?

The travels of a translocal

It is not that far from Harlow in the UK to Bali Nyonga in Cameroon. Let us follow this route with an imaginary African, who is a naturalized British citizen, setting out to return to her natal hometown in Cameroon. She is going to inspect the house she has been paying to have built using her income as a hard-working social worker in an inner London borough, but she has timed the journey to coincide with a major annual ceremony in her hometown, the *Lela* dance. The budget for the travel costs is over a thousand pounds so it is not a journey she can afford to make often. She has not been back to Bali Nyonga since her father's death celebration a decade earlier. Yet, despite her physical absence she has been in frequent contact via text messages with friends and family members in Bali Nyonga. She also regularly follows some of the internet groups that focus on Bali Nyonga affairs, so she is up to date with local news and gossip.

Harlow is a post-war new town 30 miles to the north-east of the postcolonial metropolis of London. It is one of a number of towns in the commuter belt that have become desirable sites of an escape from inner London for a generation of successful British-African professionals seeking more space for their families and the perceived moral and physical security that comes from being outside the inner city. Yet its attraction is as much its propinquity to the city as it is to its distance from it. Her feelings about both these two places are ambivalent: London is the source of her income and site of her labour, it is where her children were born and where her church is located, but she also sees it as a site of social disintegration. Through her job she has seen old people abandoned by their families, women who have been beaten by their partners and young people who have become caught up in drugs and petty crime. It is the site of the individualism that she decries and describes as 'Westernized'. Little does she know that this same epithet is frequently given to her by family members in Cameroon hoping for more remittances. Harlow, on the other hand is the place of her immediate family, where her children are being educated and where she is paying the mortgage on her house which provides her sense of security. She is happy there but she misses the close social bonds of her former life in central London. The two places are clearly part of the same process for all that they are viewed as distinct urban settlements.

After packing requested gifts and attending a send-off party from her diaspora friends, her son drives her round the north of London to the airport. Using her British passport, she then passes easily through the selectively-permeable membrane of Heathrow, with her only problem excess luggage. Early morning Tuesday finds her in Paris-Charles de Gaulle Airport, for many the door between Europe and Cameroon. As she settles in on the plane she complains to the francophone Cameroonian stranger sitting in the seat next to her about Air France's near monopoly over flights to swathes of West Africa. They bond over the absurd cost of the flight but then a scuffle breaks out at the back of the plane. Four French officials in uniform are deporting a man who keeps shouting that he is from the Democratic Republic of Congo not Cameroon. The two Cameroonian neighbours initially express pity for the man, but their tolerance wanes and shame and irritation replace pity. Their sympathy is with the man's family – the return of failure is the last thing they need just before Christmas. Once the flight has taken off, the asylum-seeker shuts up, he is willing the plane back to Paris and dreading each kilometre-click on the computer screen in front of him.

Our translocal traveller pulls the continents of Europe and Africa together into neighbourliness through the brevity of the flight to Douala and the mix of black and white passengers. The bird's-eye view cuts Europe down to size – a crowded little place on the edge of the Sahara. She has time to reflect on other journeys she has made: the nervous 19 year old who left a civil servant's household in the capital of Cameroon to study in England; the years of toil and poverty to achieve the social mobility that had reclaimed her professional status; her journey from daughter to mother and the money her father had diligently sent to her from Cameroon to help her with the early years in London; her daily commute from Harlow to London; her holiday last year visiting her sister in Texas. She thinks of those hopeful Africans way down beneath her crossing the Sahara from south to north in over-crowded pick-ups. She hopes the money that she sent for the final payments for the new house will have reached the bank account she transferred it to by the time she arrives. She looks around the cabin and wonders where the white people on board will end up tonight: oil-men, bankers, priests and anthropologists she suspects. Nine hours later she touches down in Douala, just as the day shifts to evening. Her regular phone calls to Cameroon never bring her as close to her memories of childhood as the first visceral sensation of stepping out into the Douala air.

Landing at Douala Airport any passenger feels that urgent concatenation of bodies, noise, bureaucracy, concrete, emotion, money, excitement and irritability that says 'urban'. The passengers who leave the plane are still, briefly, united by the solidarity of anxiety about immigration, Yellow Fever certificates, luggage and customs, but once through the last set of doors their brief association dissolves amidst the crowds of people waiting outside.

Douala is a city for which she has little affection. Her family home was in Yaounde, the capital, a city of politics, ministries and hills, not Douala, a flat coastal city of trade, and swamps. She has heard all the stories of crime and has no desire to linger here and so, though she is tired, our traveller continues her journey through the night. As she skirts the city centre and is driven along the portside where boats are being loaded or unloaded with commodities for or from Europe and Asia she remembers the hassle she had exporting the windows she wanted for the new house. She had bought a small minibus, filled it with European goods and the new window frames then had the doors welded shut by a Nigerian mechanic in England. But thieves on the boat had broken the windscreen and stolen the small goods. Then the 'thieves' in customs at Douala port had delayed the windows for three months waiting for a bribe from her contractor who had come to collect the vehicle. No problem. It had taken six years to build the house.

Back in the present, her car crawls across the Wouri estuary before hacking slowly through the suburbs for an hour beside a parade of tempting roadside bakeries, bars and *suya* (kebab) stalls. Like any journey through any city this transect through Douala affords our traveller only an opaque window onto the lives of the city's residents as they negotiate the opportunities behind the facade of the roadside (Ndjio 2006, 2007, Simone 2005a, 2005b,). Douala is a big city, a globally connected city, a convivial city, a fearful city, a city-in-the-making.

Winding north the vehicle picks up speed as the congestion evaporates. The main road climbs up into the hills as it moves away from the coast, passing beside (or through) some of Cameroon's significant secondary urban centres (Nkongsamba, Baffoussam, Bamenda) and through plantations of oil palm, bananas and pineapples up to the Grassfields of the North-West Province. After driving for five or six hours she drives down into Bamenda, the provincial capital where she did her secondary school education (in the same private Catholic boarding school as her mother before her).

The final leg of the journey to Bali Nyonga takes another half an hour along a newly tarred stretch of the East-West Trans-African highway, built by Chinese contractors. She marvels at the transformation that this simple technology of a waterproof road surface can make as she remembers the horror of travelling this once mudded route with her father's corpse for his burial. Now the journey is more comfortable than the crowded commute she makes daily. Never mind mobile phones or new houses, this road is the most significant material aspect of place-making in Bali Nyonga in recent years. At her journey's (temporary) end is the small town of Bali Nyonga (population c.20,000), centred on the palace of its traditional ruler (the Fon), two public squares, an old Presbyterian church, a market and a scattering of government offices housed in converted bungalows. Her chauffer has given her a local sim card for her phone and she texts her family back in England to let them know she has arrived safely.

What can we learn about translocalism from this imaginary journey? First, it draws our attention to the social relations of mobility experienced by a single person travelling through space in pursuit of their translocal goals. It draws attention not only to the banal differences of modes of transport but also to the strange synchronicities of movement – whether in a Doula suburb or a London motorway the sensation is of congestion and the equally odd synchronicities of companionability: the alliances and antagonisms that form in crowded planes or crowded commuter trains. More in line with attempts to theorize mobility (Sheller and Urry 2006, Urry 2007) it highlights both the differences and the connections between embodied, affective, material (and textual) journeys. It shows how emotions change as bodies move, but also how memories steer the emotions on any journey. How many roles does one body play as it moves? Courier of gifts, citizen, customer, carrier of practices, professional in London, elite in Cameroon, house-builder, daughter of the soil, returning prodigal, tourist, foreigner, auntie, patron, employer.

Second, the journey shows how a myopic focus on the two ends of a translocal relationship misses all the 'work' that goes on in between. This is key to the argument that the relative importance of different local places changes over time, which multiple communities mobilize in pursuit of any particular project. At each point in the journey the traveller is part of a social relationship. The house-building project is a very common form of translocal place-making for African diasporas, but it requires the builder to engage with many different groups in many different places to make it happen. The journey hints at the exponential efflorescence of consequences for these different communities that result from the core relationship of the traveller to their African home. Throughout the long history of mobility urban centres have been points of departure, sites of settlement and resettlement and sites of return, but from the perspective of individual migration it is not enormously helpful to think of translocal urbanism in terms of alliances between discrete urban spaces so much as an assemblage of urban relations. The social relations of mobility are the relations that enable mobility.

The more difficult question to ask when thinking about this journey is what does 'translocalism' add to the analysis that that 'transnational urbanism' would not have identified anyway? Clearly it succeeds in drawing attention to the fact that there are only a few moments when national identity is important. But more than that the distinctiveness of the translocal is the spotlight it directs towards the community engaged in this particular house-building project and in this particular journey. For example, this journey shows the way in which local communities are not just the outcome of movements, but the medium too – there is a recursive relationship between the practices of house-building that are physically re-making Bali Nyonga and the structures (the presence of a global Bali Nyonga diaspora, the values, the family relationships, the ideas of what constitutes 'a good life') that define the Bali Nyonga community, however widely it is scattered over space. In this sense there is scope perhaps for translocalism to by-pass familiar debates that contrast unfettered diasporas against more material accounts of obligations and social difference (Appadurai 1996a, Mitchell 1997). In some ways the traveller described here *is* an elite (though they are unlikely to be described in those terms when in the UK), who is able to finance not only the journey home but also a house they may not ever live in permanently. It is 'elites' who meant to be deterritorialized according to the literature but that is not the finding that has emerged from this research, rather it is only by engaging with the territory of 'home' that elite status in Bali Nyonga can be fully claimed. Furthermore elite status is highly relative – in relation not only to location but other factors such as genealogy, reputation, gender. Bali Nyonga is both the means of the journey as well as the destination, it is the 'how you travel' as well as the 'why you travel'. In this it is a distinct claim from that made by an analysis framed in terms of transnationalism.

Researching hometown associations: history and the question of urban 'patriotism'

The discussions in this chapter are based on a research project that looked at four small African towns (Bali and Mamfe in Cameroon and Tukuyu and Newala in Tanzania) and traced the global connections between each of them and the diasporas of domestic (national) and international migrants who had left them (Mercer et al. 2008).[1] It used a large social survey (n = 2256), interviews and observations in Cameroon and Tanzania (in the hometowns, in regional urban centres, in national urban centres) and interviews with the diaspora in cities in the UK (London, Leicester, Cambridge, Birmingham) to explore the geography and work of the 'hometown associations' connected to these four sites.

Hometown associations are clubs that diasporas formed wherever they settled in order to bring people together on the basis of an affiliation to their home area (Yenshu 2008). Given their wide geographical spread, but their narrow 'hometown' focus these groups could be the paradigmatic institutions of translocalism. Theoretically these clubs form 'chapters' or 'branches' not only in other urban centres in Cameroon or Tanzania but also wherever there are enough motivated

---

1 I would like to acknowledge the collaboration of Claire Mercer and Martin Evans on this project.

people in cities all over the world (for example in France, UK, Saudi Arabia and the USA). The aim of the associations is to provide welfare and social activities for their members in the diaspora. In addition many have also recently started to undertake development projects that aim to modernize the hometown (Page 2007). Journey's like the one described above were relatively common parts of the operation of these hometown associations. Those who travelled home from the UK to Africa not only reported back to the association on the state of the community and of the town, but also reported on the activities of the local chapters of the hometown association. Despite increasingly good communications, there is still a sense in which the power of presence and the testimony based on observation counts for more than reports on the internet or by text.

Hypothetically the different branches of the association are linked together and share information and projects. In fact such a description does not accurately capture the institutional or spatial character of these organizations. The idea that these associations are best understood as global networks focused on a single locale implies too much structure in a context where an idea of loose affiliations (but passionate loyalties) better describes the friable and intermittent quality of the linkages (Mercer et al. 2009) in which (non)-members have differential access to information, capabilities to be mobile, and varying purchasing powers. Thusfar, rather than producing a single overarching association, many places had multiple competing hometown associations (Evans 2010). The chapters within the nation were more engaged with the hometown than those overseas. In all cases the main activity of these associations is to look after their members in Douala or Dar es Salaam or Dalston (London) or Dallas and their work at home is secondary. The 'local' in the place where migrants were living was often as important as the 'local' that theoretically brought them together.

Given the focus here on the city, it is also necessary to consider the extent to which these hometowns are indeed urban. In all four case studies what is called the 'hometown' in the diaspora turns out to be a large rural district with a small town in the middle and many villages outside it. The conceptual solution of replacing a hierarchy of spatially discrete urban settlements with an analysis of a global urban process sits comfortably with contemporary urban theory. The traveller from Harlow is building a house in Bali whose architecture is styled on an American suburban villa, paid for with money from a job in London, using materials imported from factories in Nigeria and China, employing an architect from Douala, a contractor from Bamenda and labourers from Bali Nyonga. Hence it is surely reasonable to claim that whether or not Bali Nyonga achieves the status of being urban, it is part of an urban process that operates at a global scale. However, in the way that diasporas talk, ideas of village, town and city all carry distinct meanings that cannot be ignored. For our traveller from Harlow, Bali is likely to be called 'the Village', a label designed not only to signify rurality but also to signify the close bonds of a place where a dense web of connections is based on the shared history and mutual knowledge of its inhabitants. When Bali Nyonga is called the Village by those who have left to live elsewhere it might be a bit condescending in its implication of under-development, but it is also being held up by them as an ideal of community, the archetype of the local – a place where everybody knows everybody and where unchanging traditions, understood social hierarchies and close mutual support predominate. It is in this context that it makes sense to ask what we can learn about 'urban patriotism' from these four case studies?

One of the largely uncontested claims about migration is that it happens in a series of steps (Schapendonk 2010) and this has implications for identifying the places that matter to different people. Sometimes the steps in an international migration chain occur within a single generation (perhaps an individual born in the countryside moves to a secondary centre for their secondary education, a city for their first degree and then moves overseas to continue their education before becoming established in a different country) but more commonly they happen over a series of generations. Whilst the parental generation might make the move from the country to a city in Cameroon or Tanzania, it is their children (or grandchildren) who make the move between continents.[2] Our survey in Cameroon and Tanzania supported this idea showing that the people who had moved overseas had generally been brought up in the large cities of Africa rather than in rural areas.

Because international migration is generally a process of steps many international migrants have at least two towns in their home country that they can reasonably describe as their 'hometown': the town where they grew up and the town they have been taught to call their hometown because it is where their family comes from historically. Indeed it was not uncommon (particularly amongst our Cameroonian interviewees in the UK) to find people who had never physically visited their natal town, yet it was the central justification of the particular association that they joined in the UK.

The city that is most significant at an individual level is often the city that is the site of memories of childhood and family friendships. In other words this is more likely to be a big city (Yaounde, Douala or Dar es Salaam) than a small town (Bali Nyonga or Tukuyu). To some extent this is shown through ongoing loyalty to sports teams and to old schools. Yet even for those born in the big cities, rural hometowns will often also be crucial as the crucible of childhood memories, perhaps as a result of physical journeys (holidays or visits to grandparents and cousins) perhaps as a result of imaginary journeys derived from the stories of older generations. These imaginary journeys are often part of work of hometown associations. The members meet in the big cities to support each other and have fun by talking and singing in the vernacular, dancing and sharing food from 'home' and collecting money to support each other when things are bad. So a child born in Douala or Dar es Salaam *might* if their parents are active in the association be inculcated with the idea that a small town in a rural area is their family's hometown and as a result they might feel loyalty to that place, despite never visiting it. The core points here

---

2   This kind of narrative risks completely erasing earlier histories of 'international' migration connected to colonial plantation, mining and industrial enterprizes.

are that (a) there are multiple cities in Africa that emerge as simultaneous points of affiliation and (b) that the town called the 'hometown' is often not literally the hometown in the sense that it is not where someone lives most of the time.

Migration has to be put into historical perspective if these multiple loyalties are to be understood. The significance of this claim can be illustrated in at least two ways. First, the long history of hometown associations means that even though the steps in an international migration chain may be highly attenuated in temporal terms there is still a multiplicity of urban loyalties in Africa among their members. As such it is possible to conclude that for some people urban 'patriotism' is not being re-organized by the last two decades of changes usually bracketed under the shorthand of 'globalization' because such loyalties have been multiple for many years and remain so. The second observation is that government policies in relation to questions of hometown vary across time and between nations and that this can have a profound impact on urban loyalties. Again this leads to the conclusion that urban patriotisms ebb and flow in relation to dynamics other than those that are seen as particularly significant at the current moment. Both of these points will be exemplified in turn in order to articulate a broader suspicion that the transnational perspective is biased towards a claim that what happens *now* and what happens in the Global North is more significant in explaining change than what has happened in the past and what happens in the Global South. Insofar as translocalism is seen as a symptom of globalization (Appadurai 1995, Hannerz 1998) this is as much a flaw of that concept as it is of transnationalism.

Archival evidence suggests that hometown associations date back at least to the first decade of the 20th century in Tanzania and the second decade of the 20th century in Cameroon (Mercer et al. 2008: 77-130). Though their names, aims, constitutions and practices have changed over time, some of them have been in more-or-less continuous existence since then. Our own archival research suggests that even though these associations suited the goals of the colonial settlers, the initial agency that produced them was usually African not European. These were instrumental associations that enabled an ethnicized workforce to be organized, policed and communicated with. In contrast to the settlers, the colonial state was initially ambivalent about these associations because they were seen as potential breeding grounds for nationalist political parties.

After some decades of colonial rule it became clear that the model of circular migration in which African urbanized labour returned to their rural homeland once they were no longer needed by their employers was something of a fantasy (Burton 2005). Not only were plantation and mine workers becoming more settled in (or near) their places of employment, but they were also seeking to establish themselves on a permanent basis in cities. In the cities (particularly in Dar es Salaam) urban governments from the 1920s through to the 1950s sought to continue to manage African populations in tribal terms through registered tribal associations whose representatives could be summoned by urban colonial bureaucrats when there were new bye-laws or when there was trouble. Throughout the colonial period formal distinctions were maintained between (1) 'strangers' who were ultimately expected to return to their 'tribal' rural homelands and who were managed through tribal associations, (2) a modernized, detribalized clerical class and (3) indigenes (people whose 'rural' homeland happened to be co-incident with the new colonial cities). The consequence is that for around a century African families have lived their lives in urbanized capitalist contexts (cities, plantations, mines), but had associations that remind them of an alternative rural hometown. Their loyalties have been divided between places.

For example, it is quite possible that a man left Bali Nyonga in Cameroon and went to work on a plantation in Tiko on the coast in 1920 and that neither he nor his family ever returned to Bali Nyonga, and that his grandson left Cameroon for London in 1980 on a government scholarship having been picked for his academic achievements. However both the grandfather and grandson might well have been able to speak Mungaka (the vernacular of Bali Nyonga), dance the *Lo'ti* (the opening dance of any Bali Nyonga occasion and a symbol of fealty) and tell you about *Lela* (the main ceremony of the Bali Nyonga calendar) thanks to the work of the Bali Nyonga hometown association (Titanji et al 1988). Having been born in Tiko, schooled in the nearby town of Buea and gone to university in the city of Yaounde the grandson might feel loyalty to a range of urban centres, yet the hometown association he joins in London will refer to Bali Nyonga, an urban centre, where he has never lived and which he calls 'the Village'. It is important to be clear here. Only for a proportion of migrants is this scenario a possibility. For many this kind of patriotism to a rural hometown was abandoned with alacrity as they became residents in cities and left rural identities behind, however the key point in relation to the debate about translocalism is that *some* Africans were translocal in 1920, 1950 or 1980 without ever leaving Cameroon or Tanzania. It is also clear that this is a process that a focus on transnationalism tends to miss because no international border has been crossed, whereas in contrast a focus on translocalism will be more attentive to these relationships.

The second point to be made in relation to histories of urban loyalty relates to the dynamics of nation-building. Here the comparison between Cameroon and Tanzania is highly revealing. In Cameroon the project of constructing a national identity has waxed and waned, but ultimately has been profoundly undermined by the political and economic interests of the elite. In contrast the Tanzanian nation-building project has forged a more coherent identity. The difference has significant effects in shaping the way that groups in the international diaspora associate.

In Cameroon the 1960s and 70s were decades when a concerted attempt was made by the state to erase local identities. The tribal associations from the plantations were made illegal, students were sent to different parts of the country for national service, civil servants were sent to parts of their country away from their birthplaces and no rhetorical opportunity for castigating tribalism was missed. Interviews suggested that at this time it became taboo for someone to ask another person's ethnicity. Yet, in fact tribal associations had become too useful to their members to be abandoned so they just changed their names or operated underground. In addition, national service did not last long, civil servants lobbied to be relocated back home and politicians started to talk about people's responsibility to develop their hometowns. A strategy of 'regional balance' in the 1970s, which was intended to deliver spatial equality of representation, actually resulted in a hardening of regional differences and an emerging competition for the favours of the central state (Bayart 1979). Thus a paradoxical situation emerged in which the government increasingly undermined its claim to be nation-building by effectively sponsoring inter-ethnic competition (Nyamnjoh 1999).

This tendency towards sub-national segmentation along ethnic lines was dramatically exaggerated after the mid 1980s when economic stasis forced some constitutional change (Nyamnjoh and Rowlands 1998). In particular in the early 1990s multiparty politics returned for the first time in twenty years, which forced those in government to take action in order to preserve their power. Ethnic competition (orchestrated in some cases by regional hometown associations) was used by the government to undermine any effort by opposition parties to develop either an ideologically coherent programme or a united, national opposition party. This is a good example of the dangers of translocalism as a political premise. The principle used to justify this 'divide and rule' strategy was the differential rights of groups who became known as *autochthons* (the 'authentic' indigenous 'sons and daughters of the soil') and *allogenes* ('strangers' or migrants who came to the place at a later point) (Geschiere 2009, Page et al. 2010). Politics in Cameroon was increasingly organized in terms of who belonged where and an individual's political rights and responsibilities were closely associated with their ethnic 'hometowns'. It became quite normal for Cameroonians, including those in cities, to identify themselves in terms of ethnicity and to associate specifically (and sometimes almost exclusively) with co-ethnics. The introduction of ethnic categories on national identity cards only compounded this social shift.

Nowhere was this change more contentious than in large urban centres such as Douala. Here the allogenes clearly outnumbered the autochthons and, in a democratic situation, could expect to get their chosen representatives elected. But many of these 'strangers' were also often closely associated with the political opposition, whereas the autochthons were perceived to be more in line with the ruling party. People whose main urban affiliation had been to these large, cosmopolitan cities suddenly found their relationship with these places was more complicated. They were made to feel unwelcome in Douala and were effectively being told to 'return' to their own rural hometown (where they were considered autochthons) if they wanted to seek political office or social advancement via the state. For some of our interviewees these rural hometowns (which government was telling them were their home) were more or less unknown, places where they had never been, had no close relatives and for which they felt no fondness. Complex debates arose about the relative importance of the mother's and father's ethnicity in those many urban families where the parents had different ethnic identities because the marriage dated from a moment in Cameroon's history when ethnicity mattered less.

For those who did try to 'rediscover' the towns where they officially belonged they discovered they were not always welcome there either: too many generations had passed and this sudden interest represented a threat to the monopolies of local political elites and to the distribution of scarce resources such as land. Here it is possible to see, in effect, a decline in translocalism at a national scale as some urban Cameroonians found their loyalty to large cities was unwanted. This was a process fundamentally driven by *national* political interest and practice not by changing patterns of international migration. For those overseas such shifts had less traction because the Cameroonian state had less control over their personal economic position. They are content to perform national loyalty as the government seeks to appropriate their development efforts (Smith 2003) confident that the state is unlikely to impede their own ambitions for supporting their hometown.

The story in Tanzanian is very different. Here the nation-building project, though often considered rather inhumane and authoritarian in some of its policies, was far more thorough and has achieved a position where public discussions of ethnic identity remain rare and awkward. A national language (KiSwahili), the complete abolition of chieftaincy and tribal associations, the appropriation of sub-national cultural performances and ceremonies by the national government all combined towards erasing ethnic differences in the public sphere in Tanzania. This is an achievement of which many Tanzanians are both conscious and proud. It is often articulated through comparison with Kenya, Burundi or Rwanda where ethnic politics have led to violence and undermined economic growth. The success of nation-building in Tanzania is usually ascribed to the determination of Mwalimu Julius Nyerere, whose reputation in the country remains high despite a widespread dismissal of his economic policies because of the legacy of social stability. The result in Tanzanian cities is that there is far less ethnic social residential segregation than in Cameroon. Furthermore people are more likely to associate through their church or mosque than through affiliation to their hometown. To some extent this pattern was reversed by the government's endorsement of District Development Trusts – organizations in which successful urbanites come together to support education projects (such as secondary school construction) in their (or their forefather's) rural home district (Kiondo 1995). But this shift is not generating any anxiety because most Tanzanians are confident that ethnic identities have been so effectively erased over the last 40 years that capitalizing on people's 'natural' fondness for their home district poses no fundamental political risk.

Comparing these two histories of nation-building explains a fundamental difference in the way that Cameroonians and Tanzanians meet in the UK and the way translocalism works out on the ground. Whereas Cameroonians often meet in hometown associations (that is as people with a connection to Bali Nyonga), Tanzanians often meet as 'national' groups (that is people with a connection to Tanzania). So the Bali Nyonga meeting, which occurs once every two months, will draw people with an interest in Bali Nyonga from all over the UK. What unites these associations is their loyalty to an African locality. There may well be other Cameroonians living physically near to the Bali Nyonga meeting place who have their own association united by loyalty to a different town who would rather travel to the other side of London to meet their friends than go to the Bali meeting in the next street. In contrast, Tanzanian associations will bring together Tanzanians living in Slough, or Tanzanians living in Dalston regardless of the town or city from which their family comes in Tanzania. Indeed one of the dilemmas for the interventions these associations want to make in Tanzania is where in the country to make them. The character of these two forms of tranlocalism are very different. Members of the Bali Nyonga association are not only dispersed at a global scale, but they are also dispersed across the UK though they still adhere to a belief in localism. For the Tanzanians in the UK the meetings are made up of local neighbours in the UK, but with a less obvious loyalty to a particular African locality at a scale below the nation. Since members of Tanzanian associations are also geographical neighbours in the UK they are able to have more frequent and more spontaneous interactions, but in terms of their international engagement it makes much more sense to talk about their affiliation in terms of transnationalism.

This extended discussion of the histories of associational life in these two countries is intended to make several conceptual points. Firstly it shows that translocalism is not new. Since it is clear that sustaining relationships to multiple places in both material and emotional ways is a century old process, it also follows that it is possible to be translocal without crossing an international border. Second, and more significantly, the forms translocalism takes are contingent on many drivers, of which cultural and economic integration at a global scale (globalization) is only one. Very often the forms translocalism takes are driven by the policies of African governments after independence. In addition there are probably many other factors (such as those relating to resources, access, educational history, pre-colonial structures, variability in forms of status-acquisition, religious affiliation) all of which stand outside the remit of globalization and have long histories but which will shape the character of the connections between places.

Conclusions

The chapter set out to answer a series of questions about translocalism and urban patriotism in the context of a comparative study of four hometown associations with connections to Cameroon and Tanzania. The first questions concerned the concept of translocalism itself. Is there a risk that in its attempts to evade the weaknesses of transnationalism the concept of the 'local' which is integral to translocalism is being treated insufficiently critically? And, what makes it distinct?

By switching the focus away from global-local or national-national analytical frames to a local-local structure there seems little doubt that new and interesting questions for understanding the relationship between places and diasporas are generated. Furthermore for some individuals from particular places (in this case those from Cameroon) where nation-building projects have stalled it is simply more empirically convincing to give greater influence to sub-national identities than transnationalism tends to permit. But, it is rash to push nations and states too far into the background. For the Tanzanians in this study, national identity is extremely important. Even for the Cameroonians, the local place-making projects at home that are the outcome of the engagement of both individuals and associations cannot be understood without attending to the role and interventions of the state. More conceptually, it is argued that like nationalism, localism is an ideology and as such has to be considered critically.

There is good reason to be fearful about a world of local politics. Parochialism, and introversion were all in evidence in Cameroon, fostered by a national politics that emphasized autochthony. Yet it is equally absurd to say that local politics is necessarily a barrier to openness to the unassimilated other, or to solidarity at larger scales. Historically the Bali case study is one where there are established ideas about how different peoples can live together in one place and co-operate. The struggle that studies of translocalism can reveal is between whose version of history, whose portrayal of what constitutes and defines the community, whose ideas about how to treat strangers come to dominate. The claim about the dynamic, unbounded character of translocalism and places that runs through this book is empirically tenable in the context of the Cameroonian and Tanzanian case studies, but cannot be taken for granted. It takes considerable political work to keep communities open and there are many countervailing tendencies that are pushing in the opposite direction. Sometimes those tendencies come from external diasporas with their capacity for nostalgia and a melancholic desire to return places to imagined pasts.

What makes translocalism distinct is that the actions people take have a wide spatial reach because of the specifically international character of the groups being considered. In character then, translocalism is community-like (as if it were local) even within a group that happens to be dispersed over a wide geographical area. The claim of being community-like is as Sinatti (2006) has shown in her work on Senegal, normatively ambiguous in that it carries both the positive associations of mutual knowledge (leading to trust, altruism within the group and solidarity), but also the negative associations of social conservatism and inward-looking politics. The central idea of non-geographic localism is that the work that translocal migrants do is like the work that local communities do, it just happens to take a different morphology. The problem is that the concept works much better in some sites than others and that it takes myriad particular forms.

It is also worth considering at this point why the politics of hometown associations in these two African case studies are so different from the less critical accounts of those operating in Latin America, which apparently give a higher priority to economic and social development of rural areas. In large part this has to do with the political importance of autochthony (especially in Cameroon), which has meant that the associations can be drawn into the personal ambitions of political entrepreneurs. However other factors are also in play. There are fewer practical challenges when running hometown development projects in Mexico for an association based in California than there are for an association based on

Europe or the USA and trying to manage a project in sub-Saharan Africa. The relative significance of the social function of a hometown association as opposed to its development function is also related to the size, demography and wealth of its membership. However, some caution is needed here about assuming that it is the African case studies that are exceptional. Is there not scope for some more critical accounts of Latin American case studies that might start by reflecting on the celebratory discourse of migration and development?

The second set of questions respond to another one of the key themes of the book namely the relationship between theories of mobility (Urry 2007) and translocalism. What forms of mobility emerge from the translocal perspective? What are the social relations of mobility? Here the argument derives from the vignette assembled from the stories we have been told by numerous travellers moving between the UK and parts of Cameroon. Translocalism needs to address not just the two ends of a local-local connection, but numerous points in between too. For this reason it is more useful to work with a concept of a single urban process than a hierarchy of separate urban settlements. Multiple communities are mobilized in pursuit of any particular project within that urban process. Each point of the journey is marked by a series of social relationships, each of which is, therefore, part of the process of place-making within a specific local site. The translocal community is not just the outcome of movement, but it is the medium of mobility too. Exploring the dynamism and diversity of these aspects of mobility is about trying to discern which of these different points in between matters to particular people at particular times. Not only that, but if mobility and translocalism are to be brought into productive synergy then there is need to find ways to incorporate the visceral, the realm of memory, the affective landscape of the connection between locales. Translocalism is probably better at doing this than transnationalism given its more obvious connections to a wider range of emotions than just patriotism.

The third set of questions concern the significance of the urban within translocalism. Have cities become increasingly significant as sites of affiliation for international migrants? Rather underwhelmingly it seems that some cities have become increasingly significant for some international migrants at some times. Yes, there has been a rescaling of the city relative to the nation in the last two decades in Cameroon. No there has not in Tanzania. In the latter case, nation still trumps hometown as the scale of affiliation. Furthermore it is necessary to be precise about which cities have seen an elevation in importance. The cities that have increased in importance in Cameroon in terms of loyalty are probably the tier of rural hometowns, whereas the degree of loyalty to large cosmopolitan cities, such as Douala, has probably declined in recent years as a result of elite politics. All these claims need to be made with strong caveats because they only ever refer to proportions rather than totalities – there are many transnational Cameroonian families for whom Douala was and is a key site because it is both the place where home is and the place where economic enterprizes are based.

Why is it that cities are so central to producing and reproducing translocal linkages? The case studies discussed here support the claim made by Michael Peter Smith (2001, 2005a) that it is the concentration of social, financial and human resources in cities that makes them such crucial sites in projects of translocalism. In addition cities concentrate the memories and emotions that are so key to explaining the paths that people take. Cities are the meeting point of so many journeys that they will be the place where, for pragmatic reasons, communities can assemble, and through assembly can embark on place-making projects of homes. Very often the main site for the work of the Bali Nyonga hometown association was neither Bali Nyonga itself nor London, but was Yaounde, the capital city of Cameroon. This is where committee meetings were called, projects formulated, decisions made, fundraisers organized and deliberations took place. If there was a hub within the 'network' then this was it. But, again there is always the danger of fetishizing the city by giving it the capacity to make things happen to the detriment of attending to the social processes that create the linkages between and across places. From a methodological perspective cities seem the most likely sites to find translocalism at work, but this should not be mistaken for a claim about causation.

Through the *historical* analysis of urban loyalty and hometown associations in Africa presented here there is a more fundamental and distinctive point that emerges, which is that both transnationalism and translocalism tend to place too much emphasis on a connection to the recent episode of globalization. The diaspora-development nexus (lauded by the IoM (2007), the UNDP (2009) and the High Level Dialogue on Migration and Development) is based on an implicit assumption that agency lies with migrants based in the Global North who will be critical in shaping the relationship between community, city and state in the Global South (Brinkerhoff 2008, Faist 2008, Henry et al. 2004, Skeldon 2008). In some ways this echoes Michael Peter Smith's critique of Ulf Hannerz, reiterated in his chapter in this volume, that it is mistaken to only locate translocalism outside the global periphery, but adds to that the claim that it is a mistake to only locate translocalism only in the present. As part of an emerging critique of the assumptions about development and migration (Bakewell 2008, Lampert 2009, Raghuram 2009) our historical evidence suggests that such a position is implausible. Why is it that migrants suddenly become so significant and effective when they move at a global scale and when they relocate to the Global North? Why is it that movements over the last couple of decades are seen to be so important? Recent changes in technology are less important in explaining the rise of secondary towns in Cameroon than the success or failure of African nation-building strategies in the 1960s and 70s. There is, in other words a history to translocalism, which may be missed if we too quickly assume that translocalism is a symptom of globalization.

The urban landscape in Bali Nyonga is a palimpsest of the legacies of a century of migration. The buildings can be read in terms of different generations of migrants, who have brought back money and ideas that has been translated into the housing stock that is the fabric of the town. Each wave of investment brings new architectural styles and new materials to the house. Each expresses new negotiations of old social relations as different *nouveau riches* migrants engage immobile residents of the town (Ndjio 2009). Again, once these histories are related, globalization at a world scale seems somehow less coherent and less significant (Law 2004). If a translocal perspective helps us to shed light on such histories than it is doing a great service.

Chapter 9

Translocal Spatial Geographies: Multi-sited Encounters of Greek Migrants in Athens, Berlin, and New York

Anastasia Christou

Introduction: Cultural Geographies of the Greek Diaspora and the City

This chapter looks at the connections between different places and their situatedness in articulating a sense of translocality among Greek migrants in Berlin and New York, and Greek return migrants in Athens. It focuses particularly on the experience of second-generation Greeks in these three cities of the Greek diaspora. The processes of 'returning' or 'counter-diasporic' migration of second-generation Greek-Americans and Greek-Germans suggest how different places are significant. This significance of place is correlated to the nation and how identities are embedded in migrants' vision of the world and sense of territory and therefore constitute practical categories of perception and cognitive frames of reference. Migrants' experiences are shaped by physical, cultural and symbolic distance as well as proximity to the nation. By exploring the interplay of these different types of 'place sense-making' (home/host societies) we understand how migrants simultaneously perceive and act upon their relationships with cities, nations and the homelands.

This research on the ongoing social construction of the wider narratives of translocality that return migrants undertake is based on fieldwork conducted during 2007-2008 in Athens (the ancestral 'home'), and Berlin and New York ('source' migration sites for counter-diasporic movement), and explores images and experiences of place, migration and diaspora. The chapter draws on oral and written narratives collected from first- and second-generation Greek migrants as ordinary inhabitants of the city in order to address everyday urban experience and the collective sense of belongingness. The chapter explores changing images of the city as a site of belonging and/or exclusion, the strategies and identities that have influenced urban diasporic life as a cultural space of the everyday and collective efforts to negotiate a sense of intimacy in the urban locale. Throughout participants contrast the 'here' and 'there' imbued in the diasporic/counter-diasporic binary, especially members of the Greek second generation. This multi-sited research focused at exploring returnee everyday life-worlds as they unfold in translocal social spaces and thus this 'translocal fieldwork' (Smith, this volume) follows the journeys (real and imaginative) of mobile diasporic subjects in their quest for emplacement in the ancestral homeland.

As Papastergiadis notes, "critical engagement with diasporic communities in cultural studies has led to either a postmodern turn in the calls for new conceptual frameworks or to more empirical accounts of movement and settlement" (2000: 91-92). In pursuing both such aims, that is, in developing a counter-diasporic conceptual framework in this empirical study of second-generation return and settlement, I situate translocality as a central parameter in how diasporic returnees construct their sense of 'home' through place-identity and place-affiliation. Translocality then indicates that an emergent sense of 'home' in migration processes is produced through different patterns of place affiliation (and identification) that depend on life cycle changes, emotional reactions, gendered experiences etc. Thus, this view of translocality in relation to diasporic subjects challenges the view that placeness/place sense-making is static but rather eroded by acts of mobility and identification. This chapter thus unveils the multiple cultural geographies of translocal spatialities as participants narrate their lives simultaneously as 'in diaspora' and 'at home' in these different cities. Most importantly, the chapter addresses the impact of (trans)localities on how diasporic lives are lived. Diasporic city lives are saturated by socio-spatial processes through which migrant actors forge socio-cultural projects that link the 'here' and 'there' of their transnational and translocal connections (cf. Smith, this volume).

In discussing urban spatialities and temporalities of the Greek diaspora, I aim to problematize how the city as a social space is consumed, performed and contested by migrants and return migrants. This type of diaspora life in the 'home' and 'host' lands is very much a gendered, ethicized and classed process. It is also a process of numerous routine interactions such as those with the 'other' when 'migrants encounter other migrants' in the city (Christou and King 2006). Such experiences shape the ways identities and the sense of place and belonging are negotiated. More specifically, the chapter addresses the impact of (trans)localities on how diasporic lives are lived, the kinds of antagonisms and polemics that the second generation faces, as well as the degree of resistance that emerges. These geographies of resistance (Pile and Keith 1997) involving the second generation become new narratives of migrancy that 'displace' first-generation stories of hardship/struggle/success in forging a voice for the children of migrants. This action often emerges as a *renunciation* of the 'family narrative' but also as a type of *reconciliation* with the 'ideology/myth/dream' of return and a source of *agency* and *autonomy* for the second generation.

However, the agency that members of the second generation exhibit should be juxtaposed against the inherited 'culture' that filters action and shapes behaviour. Indeed, the very experience of relocation to the ancestral homeland becomes a confrontation with culture. This cultural encounter is not one that mediates 'hellenicity' (cf. Leontis 1997, Hall 2002) in known and experienced parameters but one that is localized, transposed, transformed and translated through the (post)modernized state of Greece. Return migrants are then exposed to a series of negotiations and (de)constructions of the nation that ultimately leads to a (re)definition of their own identities. In a sense, their own personal plan of action, that is, the return to the ancestral homeland as a triadic project of identification (locating the self), closure (transplanting home) and belonging (eradicating migrancy), becomes a plan unfeasible to implement, a mission impossible and a life story incomplete (Christou 2006a).

On the other hand, the *city* (both as 'home' and 'host' spaces) is also a context of action and reaction for the first generation. Here, it is interesting to note the unsuccessful attempts by the first generation to return to their homeland. That is, life in the host country becomes an attractive or necessary choice due to practical life decisions: for instance, the location there of a non-transportable ongoing family business (such as an ethnic-Greek restaurant), or financial and health care benefits that are not available in the homeland. These pragmatic circumstances create contradictory spaces of inclusion and exclusion, and of interaction and constraint, for the first generation. The narrative excerpts that are presented below intend to illustrate these interactions; and they are revealing of both 'being' and 'becoming' trajectories in the Greek diaspora.

Barkan (2004: 345), distinguishes between 'transnationalism' and 'translocalism' as the former's emphasis on simultaneity, persistence and intensity of contact/ participation across boundaries, and the latter's stress on dual social action whereupon the 'here and there' continue to exist and emotional ties persist. As an alternative to both top-down perspectives on globalization and uncritical accounts of localities as sites of resistance and hybridity, a 'transnational urbanism' (Smith 2001) therefore produces a cluster of social spaces linking individuals, communities and networks, which challenge the global/local dualism in the transnational era. By adopting a translocal approach we can go beyond the local grounding in considering broader interactions through connections and cultural affinities that shift through space and time.

My aim is to contextualize and refine the empirical study of second-generation counter-diasporic movement in relation to participant perceptions of dis/em/ placement in relocating to the ancestral homeland. I consider significant to approach counter-diasporas alongside theorizations of both immobility (the locatedness of place) and mobility (the situatedness of space) in exploring issues of migrant identification and un/belonging. I incorporate transnational conceptualizations in understanding social and cultural geographies of translocal migrancy. Such theorizations give light to aspects of 'existential migration' (Madison 2006) and the phenomenology as well as corporeality of human mobility. I seek to critically examine and situate analytic assumptions of transnationalism and translocality in conjunction with counter-diasporic theorizations. In this regard, elsewhere (King and Christou 2010) it has been discussed at great length how the literature on transnationalism and diaspora does not address the gap in focusing on the second generation in the area of return migration. In King and Christou (2010) we developed our conceptualization of counter-diasporic migration and framed this within diaspora debates and typologies while problematizing definitions of the second generation in introducing new perspectives addressing critical dimensions of mobility and return, homecoming visits and permanent settlement. We developed our view of second generation return migration as a particular migration chronotope which is an (emotional/affective, social/cultural and conscious/agentic) act of reversal of the 'scattering' (diasporization) of the first generation.

I share the perspective that diaspora and transnationalism are conceptually overlapping rather than divergent terms as they highlight not only the spatializing practices of migrant subjects but also their temporalizing practices in how migrants forge social ties within and beyond the space-time of nations (Chu 2006: 400). As Chu indicates, "the conceptual distinction between diaspora and transnationalism is more a matter of nuanced inflection than exclusive difference" (2006: 400). Here I tap into the rhythms of such a 'nuanced inflection' through this multi-sited empirical study in exploring translocal geographies of counter-diasporic subjectivities and identity projects. The concept of scale has been productively used in particular areas of geography and ecology and theories of rescaling and urban restructuring have been perceived as invaluable in the reconstitution of migration studies (Glick Shiller and Caglar 2008: 2). However, we also need to consider what Marston et al. contend by identifying fundamental weaknesses in the concept of scale and urge for the abandonment and replacement of it with alternative approaches as to "avoid the predetermination of hierarchies or boundlessness" (2005: 425). Of course this is not to suggest that the politics and policies of cityspaces as embedded in urban structures of politico-economic hierarchies are not processual in everyday migrant lives and should be abandoned. On the contrary, not only should we recognize impeding structural forces in everyday counter-diasporic lives but above all we need to consider how translocal geographies of city lives are restructured by migrants in their interplay with both local and global forces precisely through the very process of experience that shapes urban social and cultural lives of migrants. Hence we focus on this significance of locality in viewing the 'city as context' (Bretell 2003) whereupon configurations of relations take place and are shaped by globalizing /glocalizing forces, power and hegemonies.

In understanding translocal geographies of movement (diasporic) and settlement (counter-diasporic) I distinguish mobility and migration as differing phenomena and social practices (see Introduction to this volume) which "produce new subject positions, they transform and extend locality and create both new subjective experiences of place and new subjectivities" (McKay 2006a: 265). This multi-sited and multi-method interdisciplinary approach to exploring counter-diasporic encounters in the homeland and experiences of un/belonging through a translocal approach while conducting fieldwork in cityspaces of 'home' and 'away' aims to deconstruct trans/locality as a subjective process of self/other understandings. Returnee lives are situated within social and structural institutions of both 'home' and host lands yet exhibiting mobile subjectivities in trying to achieve not just physical 'grounding' but above all 'emplacement' (Smith, this volume) that will provide a new sense of homing.

The Ethnographic Urban Gaze: Visualizing Homelands and Hostlands

This section gives a glimpse of how Athens, Berlin and New York stand today as urban spaces, but also how they have been transformed as 'ethnoscapes' (Appadurai 1991). The fieldwork started in Athens, capital of Greece; the city of contradictions and complexities, that encapsulates ancient ruins and McDonalds within a single glance; the city of the 2004 Olympics; a new 'multiethnic' city; and for this research the ancestral homeland and the city of the counter-diasporic. First settled around 4000 BC, in the area that today hosts the ruins of the Acropolis, Athens is the oldest inhabited city in Europe. Its history is profoundly marked by its 'glorious' past yet stigmatized by its contemporary inadequacies. Against its ancient glory as the alleged 'cradle of democracy' (although an exclusionist space for women and slaves), Athens is nowadays a vibrant metropolis with a multicultural persona, 'home' to almost half the population of the entire country. The city's modern history is marked by political, social and economic crises, including four centuries of Ottoman control, monarchy, dictatorships, civil wars, the German occupation during World War II and subsequent famine – all markers of memory and trauma. On the other hand, Athens is a city of modernity. As host for the Olympic Games, the face-lift it received was quite impressive, adding to its infrastructure through a state-of-the-art new airport and an expanding metro service. And yet, although Greece became a member of the European Union in 1981, a sense of 'European' identity is still highly debatable; whether in cafés or their working lives, modern Athenians seem to have a different understanding of time and space, professionalism, integrity, entertainment etc. For 'native' Greeks the nation supersedes affiliation to a 'European' identity, since 'Greekness' prevails in self-identification with 'culture/civilization' and not 'topos'/locality.

Berlin was the second city of fieldwork, chosen not only because of its Greek community but because of its iconic status as a city of a reunited Europe. Lila Leontidou presents the core image of Berlin as 'the city which has become the symbol of the end of bipolarity...The Berlin wall, evocative symbol of the cold war, demolished in a global moment fixed in urban space...with cultural activities hosted by Berlin but addressed to the world' (2006: 261-262). But 9 November 1989 was not the only moment of global iconicity for Berlin. Karl Scheffler, the author of *Berlin: Ein Stadtschicksal, 1910*, stated that 'Berlin is a city condemned forever to becoming and never being'; while J.F.Kennedy, visiting Berlin in 1963, declared that he was a citizen of Berlin (*Ich bin ein Berliner*). Such statements are revealing of the aura and energy that one feels in Berlin. The new capital of Germany is a dynamic, cosmopolitan and creative city open to entertainment, recreation, science and academic life.

And finally New York: the city that never sleeps; the Big Apple; the Cosmopolis; the city forever stained by the memories of a day in history simply known as '9/11'; the city of outrageous wealth and unutterable poverty just meters apart; the city of all, everything and nothing. The birthplace of many cultural movements, the most populated city in the United States with a metropolitan area that ranks among the largest in the globe, New York has been for more than a century a major center of finance, commerce, foreign affairs, education, entertainment, fashion and arts – and immigration. Greeks were amongst the wide range of mostly European immigrants who flooded into the city during the late nineteenth and early twentieth centuries. During the great migration from the American South in the 1920s, New York City was a major destination for African Americans that brought about the flourishing of the Harlem Renaissance during the era of Prohibition that coincided with the construction of many skyscrapers forming the famous skyline of the cityscape. In the 1960s and 1970s the city suffered from economic recession and rising violence that dissipated by the 1980s with financial developments and new migration flows in the 1990s from Asia and South America. New York City is exceptionally diverse. Throughout its history the city has been a major point of entry for immigrants; the term *melting pot* was first coined to describe densely populated immigrant neighbourhoods on the Lower East Side. About 170 languages are spoken in the city. Of course, its contemporary history has been marked by the events of 11 September 2001 and the destruction of the World Trade Center. Scheduled for completion by 2012, the 'Freedom Tower' will be built on the site.

But what does it mean to look at translocal geographies of diasporic subjects' lives abroad and in the homeland from the vantage point of two distinct urban spaces, that of Berlin and New York? I argue that these two cities form a significant part of the imagination of 'globality' and 'cosmopolitanism' that transcends the mere sense of being a financial capital (New York) or a historic city (Berlin) but as sites to newly emerging transcultural and translocal relations. Here I do not see these cities as simply sociocultural categories in the world, but, as translocalities from where we can see the world, that is, through processes of globalization, transnational flows, migration, national sovereignty and citizenship. And, if we are to correlate this with a symbolic signifier, then if we take into account that Obama's presidency is vastly seen as one professing to the hope of change and a look toward the future, then his decision to make his first major international speech in Berlin overrides Berlin's previous image as the imagining of the past (in reference to the history of World War II and the Cold War) into one of the future, thus seeing Berlin as a space in the world from which to imagine the future as well as its contemporary condition, in a way similar to that of a 'changing New York' (Cordero-Guzmán et al. 2001). And, although Athens does equally 'fit' into such categorizations, it nevertheless can be seen as a hub of translocality, change and transformation given its newly emerging and vibrant multicultural city life.

Cities as sites of Counter-Diasporic Journeys

The negotiation of mobilities is a key aspect of 'counter-diasporic' journeys. In the analysis of counter-diasporic migrations from Berlin and New York to Athens, I hold that both cities as social spaces, and identities as processes of social transformation, are contexts of 'becoming' rather than entities of 'being'. Cities are sites of identifications which provide particular readings of city-spaces and participant narratives translate their changing meaning.

I understand these cities as sites of 'transnational globalization' (Smith 2001) and through the idea of the 'city as multitude' (Hardt and Negri 2000). On one level, Smith 'locates' abstract global flows in terms of smaller localities and translocal cultural and political networks. On another level, Hardt and Negri see cityspaces as locales of the multitude of peoples, nations, flows, bodies, circulations, networks and ultimately as new human geographies (2000: 396-97). Translocality recaptures the cityspace as a locus of an interactive terrain whereupon sociocultural activities are transformed and represented in diasporic lives that reflect and refract everyday encounters and structures. Such translocal cultural landscapes act as channels, either symbolically or pragmatically and so they mediate subjects' relationships with their surroundings as they contest or construct their 'sense-making' of place.

On another level, the city has been read as an entity of duality: the dual city is an ancient device, as old as the Western idea of the city itself (Cohen 2000: 318). Readings of the city as in a duality between inside and outside, material and spiritual, rich and poor, visible and invisible, bourgeois and proletarian, indigenous and immigrant, night and day run through the history of city narratives (Cupers 2005). According to Cohen (2000), the duality of the city can be related to body and text. Hence, it is important to look at migrant narratives and identity performances as both the embodiment of identification and the textual representation of urban space. On the other hand, the global city with its spaces of flows, its new ethnicities and urban lifestyles seems to deny this duality, giving birth to the image of a complex geography where inside and outside are disrupted and recombined into multiplicities (Cupers 2005). This is precisely where a translocal approach to examining cultural geographies of the city is useful in understanding urban space as a context of corporeal mobility and embodied performative settlement.

I have selected a particular way to look at the city as an urban space of 'entangled social relationships' (Sharp et al. 2000). Through the prism of stories of diasporic subjects, I view both the 'home' and 'host' city spaces through their 'cultural' narrativization of belongingness and exclusion. According to Zukin, 'culture is a powerful means of controlling cities. As a source of images and memories, it symbolizes "who belongs" in specific places' (2000: 132). Yet Zukin sees public culture as socially constructed on the micro-level, produced by social encounters and shaped by the power of culture in relation to the aesthetics of fear as accentuated by large numbers of immigrants and ethnic minorities. I aim to examine the social and cultural geographies of migrant identities in urban space and to understand the dynamics, interconnections and negotiations that take place in how cities are experienced translocally.

But first of all, through narrative excerpts I illustrate the core facets of how the city is experienced by the participants. These experiences fall under three themes.

1. Banalities of ethnic life-worlds: these experiences centre on the everyday life of migrants and how ethnic components intersect with daily routines.
2. Topographies of cultural life-worlds: these experiences sketch how culture intersects with city life in the diaspora.
3. Geographies of counter-diasporic life-worlds: these experiences focus solely on the second generation and examine the content of (re)actions in the city space during relocation to the ancestral homeland.

Having set the scene by specifying the city – above all, Athens – as a site for counter-diasporic journeys and for projects of migratory return, I now move to present some of the empirical data. This is largely in the form of selected extracts from the narrative interviews; especially the quota samples of 30 Greek-Americans and 30 Greek-Germans who were asked to narrate their life stories of return to Greece, either orally in taped interviews or in written journals.

Mobilities, Memories, Emotions: Narratives and Negotiations

The experience of migration can often become a way to channel emotions but also a hindrance to coping with the most volatile of them. Participants of the second generation who have relocated to Greece seem to experience the relocation as an irreparable rift between what they expected to find (the imaginative homeland) and what they are actually encountering upon relocation (the pragmatic homeland). More specifically, this is illustrated through the experience of Athens as the cityspace of relocation and exemplification of the national narrative of 'authenticity'/Hellenicity (where 'Greekness' and internalizations of the authentic Greek life) which does not meet the frozen in time and space image of the homeland passed on from previous generations and preserved in memory in mythologized terms and subsequently re-envisioned through translocality. In brief, this type of disillusionment and rupture occurs in relation to experiences of deep disappointment and the main axes of such are corruption, bureaucracy, access to employment, lack of concern for the environment, and the myriad frustrations of everyday life, especially in a big and often chaotic city like Athens. Furthermore, the 'pure, authentic' Greece of participants' diasporic imaginations having been fundamentally changed by globalization, EU membership and enlargement, and above all by large-scale immigration from Eastern Europe and the 'Third World' in the last twenty years, exacerbates the lack of 'purity' in its 'contamination of home' by the 'other'.

Beyond everyday life, the academic literature examining modern Greece and modern Greeks is quite consistent in presenting them both as inherently *culturally* ambivalent. This ambivalence has multiple facets: the crossroads between East and West, confusion between Europeanness and Mediterraneanness, the rapid shift from a rural to an urban lifestyle driven by mass internal migration during the 1950s and 1960s, and the juxtaposition between occidental and oriental aspects in social and cultural life reproduced through symbolic representations and life practices (Faubion 1993, Clogg 2002, Christou 2006a). According to critical ethnographic accounts (Herzfeld 1986), modern Greeks reproduce the ideological ambivalence of cultural or national identity through their symbolic uses of cultural stereotypes. Consequently, the 'other'/the migrant is an inferior being in the view of a hegemonic Greek society (Christou and King 2006) and according to a 'hierarchy of Greekness' (Triandafyllidou and Veikou 2002). In this respect there seems to be an added layer of ambiguity for the 'returnee' participants who experience xenophobic encounters and institutional disorganization that directly impact on the quality of their lives. And yet, although being traumatized by such recurrent events, they tend to (re)negotiate their 'place' and hence extend their stay in Greece. The ancestral homeland cityspace, despite its 'abusive' treatment of returnees, despite the realization that the 'mythologized' homeland does not exist, notwithstanding all this, the homeland relocation experience becomes a stamp that marks their 'personal myth'.

The 'personal myth' for participants is their personal plan of action, namely the relocation itself to the ancestral homeland as a life choice of new beginnings. Natalia, a second-generation Greek-American who 'returned' to Athens in 1977, tells us that at the time she was able to find employment at a Greek-American academic institution where she felt comfortable being with other 'hyphenated' diasporics and she experienced a large degree of unprecedented freedom as a woman who was starting her new life in her mid-twenties with the approval of her family since she was in Greece and that was (socially and culturally) acceptable. Here, it is interesting to note the contradictions that occur on a translocal level, namely in relation to the spatiotemporal context, that is, while the US in the late 70s would have been a place of freedom for young women this is not the case for most second-generationers who were deprived of such opportunities due to the strict traditional and patriarchal upbringing they had, while, on the other hand, Greece in the same era was a place of moral and ethical constraint (again in a gendered perspective) that had just acquired 'civic freedom' with the reconstitution of democracy following the military dictatorship of 1967-1974.

> I found a job immediately at the Hellenic American Union and it was a very good environment because there were about twelve of us who were hyphenated Greeks so we all had something. It was exciting being here. First of all there was also this idea of freedom – to be able to walk around being a woman, walk around any time day and night, not being bothered, you know not having a fear of you know… any kind of…violence…crime… I mean at that time Greece was perfect. So all of these things worked. And also I was in the middle of a relationship which was problematic and it was to get away and see what life would be like without this relationship. And it gave me that opportunity… so for all these reasons. And I said when I come to Greece I am going to live alone. … Was that agreeable with my parents? They said yes, fine and that gave me an opportunity to start you know at the age of twenty-four mind you, which is

late, to start figuring out who I was, what I wanted... if I could live with myself
before I, you know... because you know at twenty-four in those days a proper
lady gets married you know but I wanted to be able to live with myself first
and then figure out if I could live with somebody else you see. *(laughs)* So I
wouldn't have to put up a fight with my parents and make a breach because I
don't think that was acceptable for proper young ladies in 1977 or earlier or later.
So... and Greece was a place where I could come and stay and they would be
completely happy with that. ... So... for all these reasons.

In her narrative, Natalia also brings to the fore several core issues embodied
in the project of relocation to the ancestral homeland for the second generation.
First of all, she mentions a smooth entry into the workforce in securing a post at
an American private University which facilitated her professional development
in Greece. This is quite unusual, and unlike the many other cases of second-
generation returnees who have experienced tremendous difficulty in finding
employment despite their qualifications and experience (Christou 2006b, 2006c).
One explanation for Natalia's professional success immediately upon arrival
is most likely the fact that she relocated to Greece over three decades ago, at a
time when highly qualified applicants were scarce. Nowadays, competition for
'good' and 'respectable' jobs is fierce, due to a surplus of people with postgraduate
degrees and the few posts available. Moreover, applicants also have to compete
against those with 'connections', as nepotism is stronger than actual qualifications
and foreign-born Greeks are consequently marginalized.

Further interesting points raised by Natalia in relation to her decision to move
to the ancestral homeland are a wider sense of 'freedom' in 'walking the streets',
as well as 'walking away from a bad relationship' and escaping parental control.
These factors speak to her sense of being – as a woman seeking a new equilibrium
space to redefine herself, both as an individual with a right not to be 'afraid', and
as an individual wanting to free herself from parental and partner control.

Natalia then goes on to discuss the major changes that have taken place in the
past three decades and how these have affected her everyday life. She discusses
embodied experiences of estrangement along with cultural ones. She dwells on
feeling different as a Greek-American but also experiencing difference in the city.
Most of those experiences of alienation from her surroundings reflect *fear*:

> Yeah, I don't feel too comfortable if I walk into an area of Athens where it's a
> poor area... there are people there that look you know ... they look depressed or
> they don't look happy or they look a little you know upset...I'm scared.

Life in the city becomes a space of withdrawal and exclusion. It is a *closed*
space that comes to contrast with the sense of *freedom* initially experienced in
the conception and implementation of the relocation plan. Additional layers of
constraint appear in the daily routine of living in Athens. In a detailed account
of such experiences, Lucy also draws comparisons with Thessaloniki, having

lived in both cities. These perceptions of place through the lens of migration and mobility as an expression of diasporic subjectivities offers a spatial understanding of translocality that situates the diasporic experience within and across particular homing 'locales' (see Introduction to this volume).

More specifically, Lucy's knowledge of the city through 'engagement' with the temporal and spatial elements of place (Athens, Thessaloniki, the islands) becomes a gateway to her own sense of self. Here, we are confronted by a parallel of geographies of identities (shifting and fluid negotiations of the Greek diasporic self through everyday life experiences in the ancestral homeland) and translocal geographies (contested and ambivalent spaces of un/belonging that derive from the situatedness and connectedness between the 'here' and 'there' during relocation) that provides a more 'non-hierarchical understanding of migrant connections and networks across a range of spaces and places' (Brickell and Datta, this volume). From her narrative we are told of her realization that her 'time' is wasted, that her environmental sensitivities are confronted by wider indifference and many other issues of frustration which make it clear that life in the city in its ordinary everyday guise is far from the memories of happily relaxing vacations in Greece. Doreen Massey (2005: 151) reminds us that 'place as an ever-shifting constellation of trajectories poses the question of our thrown togetherness' and so the way we consume space does depend on what is actually 'thrown together' in a particular time. Such realizations make it clear that place is permeable not only by images of natural and built environment, but primarily by the cultural impositions that either hinder or enable its 'openness' to the possibility of living in and with the city's surroundings and practices. Such translocal reflections set the scene for a sense of belonging by the migrant as an active agent living in a time-space that is accessible beyond the imaginative. For 'relocatees' (Smith, this volume) there is indeed a transformation of the meaning of 'local' in everyday life as scale of locality extends from the macro geographies of the nation/state to the micro sociologies of home/family/neighborhood/community spaces in the city of relocation (cf. Brickell and Datta, this volume). Interactions, experiences and relationships in cityspaces are not devoid of locality and scaling as they are shaped by the type of translocal connections mentioned above. Yet, such scaling also involves a layer of emotional geographies of return and settlement as participants often dwell on the very affective aspects of situating their lives in-between the 'here' and 'there' of home/host lands in both tangible and emotional ways. Hence, the ancestral homeland return becomes a tangible interface to evaluate and refine the symbolic expression of translocal relationships through the continuous spatial constructs where 'multi-scalar imaginations' of their settlement as an attempt of emplacement (cf. Smith, this volume) reveals that diasporic subjectivities are expressions of a mission in search for grounding.

Entanglements of the Ethnos in Translocal Geographies of Return

In an interesting paper on exclusion and difference along the EU border that tackles socio-cultural markers, spatialities and mappings, Leontidou et al. (2005) refer to 'community' as a social imaginary with its attendant perceptions, spatialities and intersubjective discourses rather than as a substantive sociological 'object' or structured, cohesive whole. This reflects Cohen's (1986) argument that 'community' exists only insofar as its members maintain it through the symbols that mark its boundaries. Boundaries are viewed as either less or more permeable between those who are excluded and those who are not, and this is linked to the degree of access in crossing those boundaries. The depth, intensity and degree of permeability of boundaries depend on the way they are constructed, and on the markers utilized to define them. Leontidou et al. suggest that we must distinguish between three types of exclusion: firstly, individual exclusion identified on the basis of *social* markers; secondly, local society or 'community' exclusion identified on the basis of *cultural* and/or *ethnic* markers; thirdly, spatial/geographical/regional exclusion identified on the basis of social and/or cultural markers and inscribed in *spatialities* constructed at the local level.

The research by Leontidou et al. (2005) shows that there are many different imaginings of what constitutes exclusion, and much variation in the socio-cultural markers and mappings used to identify and distinguish it, just as there are with 'borders'. Often these deviate from the dominant perceptions of borders and exclusion. This line of analysis therefore privileges participants' concepts of what constitutes 'boundaries' and 'exclusion'. Not only do I favour such an emic approach but I further argue that *narrative scripts* are entangled with *spatial scripts* in how participants articulate their sense of self but also the degree of agency in their mobility pathways. Hence, the migrant self is scripted through imaginaries of belonging that intersect with everyday life in the city, but a city that also at the same time exposes spaces of exclusion.

In the following series of extracts, Anna, who is a second-generation Greek-German, first refers to the borders of exclusion that her mother experienced in the homeland due to her leftist ideology and activist practices in an era of authoritarianism in Greece:

> She was being followed and her life was made difficult, she was forced to leave, wanting never to return again. Even today when she narrates this incident it is a very emotional moment and she bursts into tears when talking about these stories. ...In this way I too, as a young child, experienced the reality concerning our existence as Greeks in a foreign country, the search for a cultural identity.

Anna then narrates her own search for an identity through her mother's traumatic past experiences in the homeland, marked by marginalization and surveillance due to her political beliefs. Perhaps ironically, given the family's history of exclusion,

Anna's trajectory in locating her own dual sense of self and of place is one defined by a strong pull toward Greece.

> I had a great liking to Greece, I loved Greece very much, the Greek culture which I imagined in the way I wanted to and knew about and at a certain point in time, after having lived in Germany... I experienced the two cultures intensely, I was greatly influenced I would say half and half, not more or less – I felt the powerful desire to return and so I am here today.

In a similar vein to Natalia, Anna too refers to a sense of 'freedom' that she was pursuing in defining her identity but could not 'locate' in Germany. Again, the quest for freedom is proffered as the main motivation for relocation. Anna correlates the idea of freedom with the attainment of an artistic expression found mostly in ordinary things that soothe the soul. But, she explains below, the price to pay for the seemingly abundant sense of freedom is quite enormous, as any glimpse of freedom evaporates under the immersion into everyday life in Greece. In this way the romanticized image of the ancestral homeland disappears in the midst of reality. Yet, as in the case of Greek-Americans, most Greek-Germans who have 'returned' never for a moment consider relocating back to Germany. It is surprising to say the least, that participants do not abandon the battle of adjustment in Greece, despite the challenges, for a life of relative straightforwardness and ease in Germany. Even more astounding is the fact that, both for second-generation Greek-Germans and for Greek-Americans, many of the participants relocated to Greece without their extended families, parents and siblings, who were 'left behind' in the immigrant host country.

> My spontaneity did not find any response in the German way of life and in many instances I felt that there was a restriction to my freedom which I do not know whether we could define as freedom, it is more like... it is based on an artistic psyche and I can say that this and only this was the main reason that I said that I must go to this country where I can listen to the radio, to the sound of Greece, to Greek music, to Greek voices, and at whatever price I have to pay – and I am paying it I assure you, in every aspect of my life. It is very difficult. First of all nothing functions in the same way as we are accustomed to seeing in Germany, where that which is worthy advances; on a social level nothing operates here. In all areas. In every respect. Even those things which at first seemed to be so romantic... the voices on the buses in contrast with the silence of the German way of life, the German culture – all of this is not so... things do not shine as they did at the beginning. Despite this I love them and despite this I never thought for a moment – I have been in Greece for five, six years and everyone used to say to me that for the first six years you will always have your suitcase near the door but I have not even for a moment missed... I miss my friends but I see them... even when I go to Germany to visit my sisters, soon I want to return again [to Greece].

But borders do not only exist in the ancestral homeland. Borders are experienced by those Greeks of the diaspora, hyphenated or otherwise, who would like to feel 'at home' in Greece, but who have perhaps been 'away from home' for too long to contemplate a 'real' return. This lengthy distanciation applies to many first-generation Greeks. Rebecca is one example of many returnee-participants whose parents refuse to return to Greece:

> Now you see my father never ever thought about going back, never. He lives in Germany now. I'm doing the opposite actually. He wasn't really so excited when I said I'm going to Greece. He was actually telling me, you know it's a crazy country and it's difficult and this and that and you might not be able to handle it. For him it [migration from Greece] was an escape from whatever it was he was escaping from. Seriously I'm trying to find that out and I'm actually trying to figure out, I'm escaping from Germany. I'm trying to find out what is here. Doing the opposite thing.

For both generations, *escape* seems to be part of the driving force of migration, but in different directions. Despite the attempt to escape, it seems that the homeland holds similar 'traps' for both generations. One of them being the 'othering' process that takes place for both generations. This manifests itself in how returnees are surprised by the changing face of the ancestral homeland, and also by how they themselves are perceived. For instance, in her written journal, Olympia, a 28 year-old second-generation Greek-German returnee reveals:

> When I lived in Germany I felt I was belonging to this cultural group. As I'm living now in Greece I don't feel it. I'm aware of the heritage and I try to respect it in any way (as for example in not throwing litter into the sea, spending money for renovation of the temples etc.) and I see that today's Greeks behave more like uncultivated, uncivilized bush-people. I don't identify myself with these. I was proud in Germany, boasting that I was a Greek, especially because of the heritage, but here in Greece, I feel ashamed of saying I'm Greek. If you look around and see what 'they' have done to Greece... it makes you feel ashamed, not proud... Unfortunately, Greece has changed in my point of view from bad to worse. Although it is since 1981 in the European Union, it hasn't developed enough. Neither economically nor socially. I think that the Greeks have transformed in a kind of 'pseudo-Europeans', pretending of having a high standard of living, by driving expensive cars and living in high-society neighborhoods, but actually the majority of these are in debt. I noticed that the friendliness of the Greeks vanished. Everyone is looking to get through the day, without considering the people around. I feel that I have changed a lot, too. I'm not that patient anymore and I am tough with people around me. I do not trust anyone and I turned to be very egoistic. This happened since I first arrived here.

It is a bit clearer to understand that a hyphenated diasporic Greek can become the 'other' through the cultural hybridization that mostly defines second-generation identities (Christou 2006a). Yet it is still surprising, to an extent, how easy it is for the diasporic first generation to become stigmatized as the 'other'. Here is a dialogue between two first-generation Greek-German social workers interviewed in Berlin:

> Fotini: A type of complaint that the Greeks who are living abroad still continue to have is the fact that when they return to Greece they face problems because people say, 'Ah! The German lady came' or 'the woman from Berlin' or I don't know what...People do not embrace them as they should. And all of this... that is the psychology which is generated around the immigrant is made up of and is influenced by all of these...these dynamics.

> Antonia: These dynamics exactly. And this psychology is what all of us carry with us more or less. Isn't that so? The immigrants, the immigrant workers, the simple immigrants and also each and every one of us has been influenced by this psychology. We would not have the same psychology if we had remained in Greece. It is these conditions which define in a way what our future and what our present is like.

> Fotini: And of course a Greek living abroad for over 30 years acquires a different identity. That is no matter how Greek he [sic] feels and even though he has not assimilated in the foreign society, he picks up elements of that foreign society. In other words, his identity is diversified in a way and this is the problem which most Greeks who return to Greece face. This is especially true for the first-generation immigrants who believe that they will find Greece as it was ... This Greece does not exist anymore because the Greeks have developed, they have changed. I mean those Greeks who reside in Greece. And this creates a new problem. I see this with the... with my own self when I go down to Greece.

In discussing identities and relocation projects there seems to be an abundance of prisms and kaleidoscopes. The very experience of migration is such a process, involving not only new experiences but also new windows, new views and new outlooks into the seemingly similar but inherently different homeland and its people. In this sense we understand that the self is divided at home and in the diaspora and that space is not a passive container of identification but a dynamic *translocal site* of (intra-) and (inter-)subjective cultural encounters. Neither the individual migrant nor the collective 'community' can be accommodated if narratives of identity are not re-imagined and if the 'community' is not reconfigured to encompass the possibility of the 'other' becoming one of our 'own'. Then, if the boundaries require blurring to reflect blurred identities, in that case the homeland needs to be reconceptualized not as a borderland but as a boundary-homeland. This leads to the realization that 'the borderlands of family and culture provide space for overlapping and complicating life circumstances to produce more elaborated and flexible models of individual development and family structure, and ones which are appropriate to the globalized and postmodern self' (Souter and Raja 2008: 26). Ultimately the ambiguous view of 'home' signifies that 'homecoming' is not a static state of 'being' but a fluid process of 'becoming'. Homecoming then is but a journeying into city-spaces of a self floating into, but also out of, the translocal boundaries of the homeland.

Conclusion

In this chapter I have tried to explore new ways in which urban spaces become appropriated, mediated and negotiated by migrants, especially second-generation Greeks returning to their ancestral homeland by relocating to Athens. Using a cross-section of the research material – narrative interviews, discussion groups and written accounts – I have shown how cityscapes are dynamic sites of complex relations wherein life stories are forged and identities develop. My reading of such spaces is informed by migrants' accounts of their counter-diasporic journeys; such narratives are spatial scripts of cultural mobility. The narrative excerpts that I have presented reveal that urban translocations of Greek diasporic life confront spaces of power, control and fear, as well as of comfort, liberation and belonging. The cityspaces of the ancestral homeland are not always welcoming spaces; rather, they seem more often to become locales of exclusion and disappointment. Herein lies a fundamental contradiction which has been discussed elsewhere (Christou 2006c) in how return 'dreams' materialize into 'nightmares', as well as in several chapters of Markowitz and Stefansson (2005) in exploring 'homecomings' as 'unsettling paths of return'.

For the most part, returnees have forged a plan of action, the essence of which is their romantic myth of the homeland return and immersion into the ethnocultural world of their ancestors and Greece-based relatives. However, the romanticized image of 'home' disappears when they are confronted by the reality of everyday life in the new, fast-changing, 'modernized' Greece. The complexity of their situation, and its contradictory nature, arises from the fact that, although the romantic myth dissipates under the welter of tough experiences upon return, this confrontation with reality does not substantially demolish the power of the narrative, or the personal plan of action, which continues to be 'lived out'. This is a case of *narrative override*. Whilst the returnees can see in the mundane world of their everyday experiences that living in Greece is no cake-walk, and in fact is in severe contradiction to their romanticized myth, nevertheless the narrative continues to be a 'commanding voice' in the way that they go on living their lives. It is almost as if their narrative, and their psychological attachment to it, transcends their troubled encounters with reality, and so they continue, whether they are Greek-Americans or Greek-Germans, to go on battling for their narrative

– and physical – return to Greece. This central contradiction is the essence of the outcome of their translocal and (counter-)diasporic lives.

Acknowledgements

An earlier version of this chapter was presented at the 2008 *Association of American Geographers Annual Meeting*. I am grateful to the Department of Geography, University of Sussex for providing the funding to attend and make the presentation. I gratefully acknowledge the financial backing of the AHRC (grant no. AH/E508601X/1) and the support of my project team colleagues. Special thanks go Dr D. Mentzeniotis for inspiration and encouragement. As always, my heartfelt gratitude goes to all my participants who have so generously devoted their time and energy to my research and embraced me and the project into their mobile lives.

Chapter 10

Translocality in Washington, D.C. and Addis Ababa: Spaces and Linkages of the Ethiopian Diaspora in Two Capital Cities

Elizabeth Chacko

Introduction

In this chapter I make a case for cities as important spaces where translocal (local to local) and transnational/diasporic linkages are formed, transferred and reinforced. Using the case of the Ethiopian immigrant community in the United States, I demonstrate how identities, histories, imaginaries and communities span scales that are simultaneously hierarchical (for example, city, region, country) and lateral (between neighbourhoods, between cities and between countries), but are strongly grounded in real space/place. Specifically, this chapter examines the nature and significance of translocal connections forged within and between metropolitan Washington, D.C. and the city of Addis Ababa by the Ethiopian diaspora in the United States. Translocality is used as a lens to understand the process of international migration from Ethiopia to the United States, the internal migration of Ethiopians within the United States, the formation of spaces for the community to meet and lay claim to, as well as the flows of financial capital and assistance from the Ethiopian diaspora in the United States to the home country.

The idea of locality and place is sometimes considered fragile in an era of increasing mobilities as linkages are forged and maintained across boundaries. Appadurai (1996a) in defining the term translocal, suggested that such linkages, be they related to labour, trade or cash flows; media or travel should not only be given primacy over place, but asserts that place is becoming irrelevant as a factor in forming community. In this paradigm, translocality includes the institutional but not the geographical. Immigration and other mobilities across and within boundaries have resulted in an emphasis on translocality as "the sum of linkages and connections between places (media, travel, labor, import/ export, etc." (Mandaville, 1999: 672) while abjuring the notion that translocal ties are related to these very places.

It is indeed true that individuals and households are linked to one another via immigrant, ethnic and other networks that increase the migrants' access to social, financial, cultural, and informational resources. However, flows of people, ideas and commodities occur in space-time and the networks formed are accompanied by processes that often reify place making and place differentiation. Indeed, the case of the Ethiopians in DC supports Michael Peter Smith's (2005a) oppositional argument that cities are vital sites of transnational ties that link people and places across the world and that there is a strong sense of commitment and locatedness within the local context in urban areas. According to Smith, cities and localities provide prospects as well as restrictions within which transnational actors and networks operate.

The relationship between locality and translocality for diasporic groups such as the Ethiopians can be understood in terms of mobilities of persons, flows of culture, capital, ideas and technologies between urban places and these relationships are implicated in the restructuring of local cultures, communities and economies. Diasporic Ethiopians have simultaneous attachments to multiple localized places such as the cities of Washington, D.C. and Addis Ababa and to specific neighbourhoods or blocks within these cities, as well as to particular educational and religious institutions located in different locales. Neighbourhoods and localities are sites where immigrant identities are emplaced even in the highly diverse Washington metropolitan area. The Ethiopian immigrants' strategies of organizing and mobilization described in this chapter also point to both an attachment to particular cities and places within them, as well as agency on the part of the immigrants in their quest for creating localized ethnic space in an immigrant gateway and making an impact in Addis Ababa, the capital of their country of origin. Immigrants' actions reflect simultaneously the group's attachment to material and symbolic spaces in the cities of Washington, D.C. and Addis Ababa as well as the national and global reach of their translocal connections.

The findings presented in this chapter draw on over seven years of on-going research related to the Ethiopian immigrant community in the United States. I use information obtained through key informant interviews with local community leaders who spearheaded the push to create a "Little Ethiopia" in Washington, D.C. and through participant observation at various community events. I also interviewed first and second-generation Ethiopian immigrants with a view to understanding how the community stayed connected and how individual ethnic, racial and immigrant identities evolved over time. I conducted focus group discussions with Ethiopian immigrants who had started businesses in the Washington metropolitan area and interviewed Ethiopian immigrant entrepreneurs who had returned to Ethiopia to start business ventures there. Additionally, I obtained data on investments in Ethiopia by the diaspora from the Ethiopian Investment Agency, an Ethiopian government organization located in Addis Ababa whose function is to facilitate and monitor investments from abroad. As I map the contours of the community's translocal and diasporic connections in the following pages, I demonstrate that far from being deterritorialized, the Ethiopian diaspora is rooted and connected in cities and that urban space/place and capital cities in particular have great symbolic significance for this group.

Migration and translocality in the Washington metropolitan area

The Washington metropolitan area, not considered a traditional immigrant city, has seen a burgeoning of its immigrant population since the 1970s (Singer, 2004) and is now considered one of the top immigrant destinations in the United States (Price and Singer, 2008). According to the U.S. Census 2000, Metropolitan Washington had some 832,000 foreign born, while the 2006 American Community Survey estimates the figure at over a million, with 1 in 5 residents being foreign born (U.S. Census Bureau, 2000; 2006). Immigrants from nearly 200 countries have settled in the central city of Washington, D.C. and its suburbs within neighbourhoods that are organized by socio-economic status rather than ethnicity or nationality (Price et. al., 2005). Ethiopians, who are among the Washington area's top ten immigrant groups, are also the largest group from sub-Saharan Africa in the metropolitan area (Chacko, 2008).

Although driven primarily by economic and political factors, migration is also influenced by experiences, as well as perceptions and views (formulated, as well as internalized) of areas of origin and possible destinations. Elements of popular imaginary related to cities, such as iconography, oral narratives, and visual symbols get collectively defined and redefined over time resulting in new forms of social imaginaries, with multiple meanings and values. Capital cities, urban tourist destinations and iconic cities are often imagined and formulated in particular ways in the minds of their dwellers, the citizens of the states in which they are located and those living elsewhere (Lewinson, 2003; Nagel, 2002; Neill, 2005).

Washington, D.C. holds a high status in the minds of emigrating Ethiopians as the capital of the United States and a city of global importance. To them, it is urban space that reflects and transmits the strength of the world's only remaining superpower (Focus group discussion, 2007). It is therefore not surprising that Washington is the city with the largest Ethiopian population in the United States, housing nearly a fifth (~17 percent) of the entire reported population of Ethiopians living in the United States. The region also is estimated to have the world's biggest Ethiopian community outside of Africa, according to the Ethiopian Embassy. American Community Survey data indicate that there are 27,703 ethnic Ethiopians in the Washington area (U.S. Census Bureau, 2006). But the Ethiopian Embassy and Ethiopian community leaders in the city argue that the Census figure is too low. Reported numbers for the community, depending on the source can vary between 100,000 and 200,000.

Ethiopians have been drawn to the Washington area since the 1960s. Early members of the local diasporic community were largely composed of students, members of the diplomatic corps and a few professionals who belonged to the educated elite. Some members of this small group stayed on in the United States as they acquired work visas, permanent residency and finally, citizenship, forming the nucleus around which a growing Ethiopian population was built (Selassie, 1996). The 1974-75 civil war in Ethiopia, and continued violence under dictator Mengistu Haile Mariam's government spurred outmigration during the 1970s, while the 1980s and 1990s saw a large flow of refugees and asylum seekers (most belonging to the Amhara ethnic group). They sought to escape persecution and war and could enter the United States as political immigrants following the passing of the Refugee Act of 1980. The community continued to grow in Washington as a result of ongoing political and economic turmoil in Ethiopia and more lenient immigration legislations in the United States that favoured family unification, offered refuge to persons at risk of persecution and promoted diversity of its new immigrant populations.

Like other immigrants to the United States, Ethiopians tended to gravitate towards cities with an established community of compatriots and many were drawn to the Washington area with its well-known Ethiopian community. The early flows of immigrants, refugees and asylum seekers from Ethiopia to the United States while structured and constrained by U.S. laws and acts pertaining to immigration, were also influenced by translocal imaginaries of the favoured destination, Washington, D.C. Cartier (2006: 138) posits that "actual person mobility" is supplemented by "imaginaries about such possibilities of mobility". Such imaginaries were constructed by both Ethiopians residing in the U.S. capital and potential Ethiopian immigrants to the United States. Through letters, phone calls, as well as photographs, and narratives, representations of the city as well as information on its Ethiopian community and ethnic institutions were communicated to Ethiopia. In addition to the importance given to the Washington area as the locus of demographic concentration of the Ethiopian population, the metropolis, the central city of Washington, D.C. and neighbourhoods within it were evoked through images of a home for the community, social bonding in physically rooted places such as churches, ethnic stores and restaurants, and reinforced by translocal flows of immigrants on return visits to Ethiopia.

Although not a traditional immigrant city, by the first decade of the 21st century, the Washington metropolitan area was characterized as a new immigrant gateway with accompanying diversity in ethnicity and nationality among its denizens. The metropolis' new settlers demonstrated much less residential clustering by ethnic/national group than their counterparts who settled in ethnic enclaves (a form of localization) in U.S. cities in the late 19th and early 20th centuries. The new immigrants' patterns of residential location and community formation can be described as heterolocalism, which refers to "recent populations of shared ethnic identity which enter an area from distant sources, then quickly adopt a dispersed pattern of residential location, all the while managing to remain cohesive through a variety of means" (Zelinsky and Lee, 1998: 281).

The Washington metropolitan area is not a major port of entry for Ethiopians, but is an important destination for secondary migration streams; immigrants, refugees and asylees who moved from the areas and cities in which they first lived or were settled by the Office of Refugee Resettlement on arrival in the United States. As these secondary movements took place and Washington grew as an area of Ethiopian settlement, local, regional and national affinities formed and strengthened in Ethiopia were transferred to networks and spaces in the

Washington metropolis and its neighbourhoods. Once in Washington, D.C., Ethiopian immigrants' ethnic spatial activities were partitioned on the basis of networks related to religion and religious denomination and social groups related to city or town of origin, familial and ethnic (tribal) ties, educational institutions in Ethiopia, political affiliations and socio-economic status, but the group collectively envisioned Washington, D.C. as an "Ethiopian city". Often, the reasons provided for relocation to the Washington area were related to the presence of a large number of co-ethnics as well as an abundance of ethnic institutions and services as the following example demonstrates.

Michael, now the co-owner of a thriving Ethiopian restaurant in Washington, D.C.'s Shaw neighbourhood reported that he felt like he was in his own country when he moved from Addis Ababa to Washington, D.C. in the 1990s. Michael's father had decided that it would be best for his son to complete his high school education in the United States and sent him to stay with his uncle in D.C. The young man attended D.C. Public Schools and although he faced multiple adjustment problems at school, the presence of family, a large Ethiopian community and particularly young ethnic Ethiopians in the city with whom he bonded and made friends eased his transition. Michael found it relatively easy to connect with young Ethiopian immigrants in the city through its local ethnic institutions and through family networks. He remarked, "There really wasn't that much of a difference in the way we thought or operated. We were friends and brothers."

Secondary migration is exemplified in the case of Yodit, an engineer who used to live with her husband in Bloomington, Indiana and moved to the city of Rockville, Maryland in suburban Washington. This is how she explained the underlying motives for their re-location:

> Both Eyob (her husband) and I got our degrees from American universities and both of us got good jobs in Bloomington soon after we graduated. But when we were thinking of starting a family, we knew it was important to be close to our people. We wanted our kids to be around people like them. To see Ethiopians, to have Ethiopian friends, to be a part of family and community celebrations. Eyob has an uncle in Baltimore; I have cousins in the DC area. So really, we came here for family, for community. Eyob found a job here before we moved, but I didn't. But no matter. I started my own business and it was hard work, but I think it's paying off. I am glad to be around other Habashas[1] and glad for my kids' sake. We are not isolated anymore.

Ephraim, who moved to the Washington area from Boston, had a similar story. His high school friend, who lived in Alexandria, Virginia in metropolitan Washington, persuaded him to move:

---

1   Usually, a person belonging to Ethiopia's Amhara or Tigrinya ethnic groups.

...so it's not like there aren't Ethiopians in Boston; there are. But when I visited DC to see my friend, I thought, I like it here. There are lots of Ethiopians, lots of Africans. Better weather too! I think I could fit in pretty well. There are a bunch of guys from our high school here in Washington. My friend encouraged me. He said I could stay with him and he would ask around for a job for me. I could not have moved here without my friend's help. I started working in a parking garage and now I drive a taxi. I will never forget my friend's help. I owe him a lot.

The geography of community formation in the context of international migration from Ethiopia to the United States as well as the internal movement of the immigrants between states and cities is affected by knowledge and perceptions of the "local" (cities, neighbourhoods) in Washington conveyed to them through kin and friend networks, through images transmitted by the media and through personal experience. These migrants in turn affect the "local" by adding to the pool of Ethiopians who increasingly impact the city and its neighbourhoods. As immigrants settle in cities, they change the demographic, cultural, economic and political make up of these urban places. Cities and city neighbourhoods besides occupying material space are also social spaces, evolving entities that are places, which are defined by Horvath (2004) as "... junctures of distinctive practices that arise particularly at times of change". The Ethiopian presence in the Washington metropolitan area is palpable, as Ethiopian immigrants enter and dominate different occupations such as taxi cab driving or parking garage attendants. They also make their presence felt through ethnic restaurants, cafés and stores and through increasing political activity and activism. It is therefore not much of an exaggeration when Washington, D.C. is referred to as an "Ethiopian City".

Translocal urban space, identity and politics

In an age of heterolocalism, how do ethnic communities such as the Ethiopians construct and express their identities in urban space? In the face of increasingly de-territorialized linkages that bind a community, are immigrants making efforts to re-territorialize identity through ethnic place-making at the level of the neighbourhood and city? Are they using locality-based networks in different places to effect political changes that could have ramifications at regional and even national levels?

According to Zelinsky and Lee (1998) dispersed ethnic groups form community by gathering in ethnic churches, business associations, athletic leagues, social service clubs, bars, cultural festivals, and institutions, which are not usually located in areas where the group also has a higher level of residential clustering. However, there is evidence that even immigrants with resources, who tend to locate increasingly in dispersed settings, are likely to establish residences near friends and relatives (Allen and Turner, 1996). Even limited residential and commercial clustering in cities, besides facilitating greater community social interactions can help create a rudimentary ethnic sense of place. Clustering can also help maximize the group's political voice, especially in highly diverse cities. The concentration of co-ethnics has been identified as important if a community is to acquire greater visibility and political influence within a city (Clark and Morrison, 1995).

For Ethiopian immigrants, the attraction of Washington, D.C. as a place was multi-faceted. In addition to being a relatively large urban center that offered economic, social and cultural opportunities, it was the U.S. city that had the most Ethiopians. Despite the emergence of an array of ethnic Ethiopian cultural, social and political associations, secular and religious institutions, as well as restaurants and cafes that were dispersed across the Washington metropolitan area, with clustering of Ethiopian businesses in many localities (Chacko, 2003), Ethiopian immigrants desired to develop an official ethnic enclave in geographically bounded space. They wished to create a place of ethnic pride in the city of Washington, D.C. that would have city recognition and would be marked via street signs and labels on city maps as "Little Ethiopia" (Chacko, 2008).

The symbolic status of Washington, D.C. within the United States as "the nation's capital" and to the rest of the world as an important seat of global power carried the greatest weight in the immigrants' deliberations and desire to claim ethnic space in Washington, D.C. In their quest for an urban enclave that would clearly bear the geographic expression of their ethno-national identity, Ethiopian community leaders were clear that they desired the official enclave to be located in Washington, D.C. Even though a more concentrated Ethiopian population existed in Alexandria/Arlington in neighbouring Virginia, members of the community believed that locating the enclave in the nation's capital provided prestige and recognition in a manner that a suburban location would not (personal communications with community leaders).

To Ethiopian community leaders, the capital and the nation were not just connected; in their minds the two were conflated: Washington, D.C., the nucleus of national and international political power with its assemblage of influential institutions and visibility on the world stage was the city that represented the might of the United States. It was regarded by the immigrants as the hub of a global network of political and socio-cultural ties, instrumental in forging connections between nations, cities and organizations with transnational reach. Washington, D.C., to them, was the urban face of American superpowerdom and hence the city in which they believed they could showcase ethnic presence and pride best through spatialization and the formation of an official ethnic enclave.

In late 2004 Ethiopian immigrants mounted efforts to designate a section of 9th street off the U Street thoroughfare in the Shaw neighbourhood in Ward 1 in Washington, DC as "Little Ethiopia". Ward 1, the most diverse ward in the city, housed less than 2 percent of Ethiopians living in the metropolis, but boasted over 40 restaurants and services run by members of the community. Ethiopian immigrants with businesses in the Shaw neighbourhood prided themselves on being active participants in the revitalization of this once run-down area (Chacko, 2008). In 2001, Ethiopians in Los Angeles had been successful in getting the city to authorize the re-naming of an area in LA as "Little Ethiopia" (Chacko and Cheung, 2003) and wished to replicate that success in Washington, D.C.

Clearly, Ethiopians were strategic in choosing both the neighbourhood and the city in which they wanted to have a second official ethnic enclave. The significance of having officially marked ethnic space in the country's capital and the symbolic heft of such an action was important to these immigrants. This desire to claim space in a hegemonic center is not limited to immigrants in the United States. Jacobs (1996) using the example of London, describes how power structures that gave rise to Empire and imperial cities live on in the negotiation of power and identity in the contemporary city. Through a case study of the formation of a Banglatown in a London neighbourhood, she shows how Bangladeshi immigrants used essentialized notions of their culture to control development in favour of local immigrant businesses and in creating a recognized local ethnic enclave.

In their efforts to form a Little Ethiopia in D.C. Ethiopian immigrants mobilized the community using multiple strategies and through different media. The mission of organizations such as Ethiopian American Constituency Foundation (EACF), established in 2003 in Washington, D.C., is to facilitate the involvement of Ethiopian immigrants in U.S. political practice and acquire greater visibility and political influence for the group. Immigrant leaders mobilized supporters of "Little Ethiopia" via an online petition run by the EACF and garnered several thousand electronic signatures championing the demarcation of ethnic Ethiopian space in the city. The flexibility and elasticity of a translocal Ethiopian community were put to use as proponents of "Little Ethiopia" reached out to advocates from across the nation, and indeed, from across the world. Individuals forwarded the petition to relatives and Ethiopian friends urging them to vote online and add their names to the list of supporters of an Ethiopian ethnic enclave. Although the formation of an official "Little Ethiopia" in Shaw did not materialize due to resistance from native African Americans who comprized about 80 percent of the population residing in the neighbourhood, the online petition was successful in that it triggered the reconstituting of the Washington Ethiopian community into one that transcended local, regional and national boundaries.

However, as Soja (1989) reminds us, people make places and places make people, resulting in 'sociospatial dialectic'. It is pertinent to remember that the online Ethiopian translocal community described here, while based on a web of relational connections and formed in virtual space with the assistance of modern technologies, was created for the express purpose of permanently delineating ethnic community in physical space. A translocal network that was capable of stretching to include people who were miles and even continents apart was tied to the demarcation of a section of a street in a north-west DC neighbourhood with the goal of enhancing a sense of belonging, pride and home for a dislocated/displaced Ethiopian community. Moreover, both the organization that spearheaded the online petition and the persons who voted were anchored in localities and initially connected through place-based linkages. The understandings and imaginings of a translocal community of what it meant to fix ethnic identity in material city space informed the actions of proponents of Little Ethiopia.

The immigrant community also engaged with local politicians to obtain assistance in creating differentiated urban place that symbolized the recent history, economic and cultural imprints and ethnicity of Ethiopians in the city and in the neighbourhood of Shaw. Ethiopian leaders enlisted the support of Ward 1 council member Jim Graham in their mission to create a "Little Ethiopia" in the U.S. capital. Ethiopians in DC often referred to Councilman Graham as their "mayor", noting that he was very responsive to their needs. Considered pro-immigrant, Graham was responsible for helping pass a law that required District of Columbia government agencies to have bilingual employees and to translate vital documents such as applications and complaints forms into Spanish, Mandarin Chinese, Korean, Vietnamese and Amharic, the last being the primary language of Ethiopian immigrants in the metropolitan area. In response to questions about his support of the "Little Ethiopia" initiative, Graham reportedly praised the Ethiopian community for its significant contributions to the Shaw neighbourhood by opening restaurants, hair salons, churches and a community services center (Westley, 2006).

In 2004, the community in turn arranged for Graham and his aide to visit Addis Ababa, Ethiopia's capital city. The rationale for the trip was for Graham to understand more about the culture of this group of immigrants in his ward and to serve his constituents better (Aarons, 2004). The Ethiopian community in the Washington area continued to maintain a strong relationship with the councilman. In 2007, the year of the Ethiopian Millennium, Jim Graham was presented with a Certificate of Excellence Award for being a "life time friend of the Ethiopian community" and for his contributions to the advancement of this group of immigrants. In 2008, he was a special guest at the Ethiopian Sports and Cultural Festival held in Washington, D.C., the only non-Ethiopian guest of honour.

Although Ethiopians only formed a very small proportion (2%) of residents of Ward 1 (U.S. Census Bureau, 2000), their business interests in the area made it vital for them to be aware of local politics and planning and allocation decisions. While individual Ethiopians may have had little stake in DC politics, collectively the group sought to influence laws and decisions that would benefit their compatriots whose businesses were clustered in the U Street neighbourhood and elsewhere in the city while raising their visibility as a group and bringing about a deeper engagement with the city. Given the heterolocal nature of their residential settlement across the metropolitan area, it is evident that Ethiopians' strategies made the best use of the localized nature of their business cluster around U Street in D.C. and the translocal connections between the commercial cluster and the scattered community that converged on this incipient ethnic enclave periodically to meet each other in the neighbourhood's Ethiopian eateries or use the ethnic institutions and services located here. These translocal ties served to support the notion of Washington, D.C. as an "Ethiopian city".

Translocality illuminates the many processes and scales at which mobilities occur, blurring boundaries between the local and the global and between local, national and global/universal. Political actions and claims made at local or national levels can "jump scale" (N. Smith, 2008), to gain credence as global or universal concerns. The translocal connections of Ethiopian immigrants were extended across national boundaries to link the capital cities of Washington, D.C. and Addis Ababa. By obtaining the support of a local politician whose jurisdiction was limited to a section of the city, but whose visibility and influence were city-wide and by taking him to Addis Ababa and introducing him to local politicians there, Ethiopians attempted to solidify their political connections and strengthen their platform at multiple scales. Through these strategies, they brought more focused attention to their activities within a D.C. neighbourhood and the city of Washington, D.C. while providing Councilman Graham an avenue to raise his profile both nationally and internationally. Local neighbourhood level mobilizations for official ethnic space were in this instance also combined with issues of creating a political presence for the group in the city. The fluidity of translocality can be seen in the manner in which the local Ethiopian community successfully "jumped scale" from the local to the national/global levels through the use of a key local politician. They transformed a process that would usually be conducted through formal diplomatic procedures into one that was both transnational and translocal.

Translocal connections – Washington and Addis Ababa

Addis Ababa, Ethiopia's capital city since the late 19th century is a primate city with a population of 2.74 million (Central Statistical Agency, 2007). The largest urban center in the country, it is more than ten times the size of the next largest Ethiopian city, Nazret. Addis Ababa is sometimes referred to as the political capital of Africa; it is here that the headquarters of the U.N Economic Commission for Africa and the African Union are located. Addis' pan-African political and diplomatic significance and its antecedents as an indigenous rather than colonial city is a source of great pride to Ethiopians.

In Ethiopia, the construction of identity takes place primarily along the lines of religion, religious denomination and ethnic group, but with considerable emphasis on geography and territory, which are often used as markers of ethnicity and difference. Ethiopia is divided into nine administrative regions based on major ethnic group, and in addition has three chartered cities (Addis Ababa, Harari and Dire Diwa). Although the majority of Addis Ababa's population comprizes of persons of Amhara, Oromo and Gurage heritage, it is a cosmopolitan and multi-ethnic city, housing people of over 80 nationalities (Central Statistical Agency, 2007).

Despite an Ethiopian government policy that discourages the free movement of its population across intra-state provincial borders, Ethiopia's capital continues to draw migrants from across the country. Those living in the smaller urban centers of Ethiopia often desire to move to Addis Ababa, which they see as a place of opportunity and likely upward socio-economic mobility. Parents hope to send their children to the capital to pursue education or job opportunities. Particularly for those from outside the capital, living and working in Addis Ababa and the prospect of having access to the myriad opportunities it offers is often conceived of as the pinnacle of place attainment within the country.

An immigrant who hailed from the city of Mek'ele in the northern region of Tigray reported that it was his aspiration as a child to study at Addis Ababa University and live in his country's capital:

> Anyone with ambition wants to go to Addis. I dreamed of going there, living there, studying there. Thanks to God, I was able to do these things. I studied urban and regional planning at Addis Ababa University and got a job in the government. I would not be here [Washington, D.C] if I had not gone to Addis, I think.

U.S.-based Ethiopian immigrants maintain their relationship with the home country and its capital and are playing an increasingly important role in changing the home country's economic landscapes. Not only are they the group that invest the most in Ethiopia and Addis Ababa in particular, but also tend to be the most politically active and influential of any Ethiopian diasporic group. Since the mid-1990s, the Ethiopian government has devized policies and provided various incentives to attract foreign direct investment and involve its diaspora in economic development. The Ethiopian government enacted a law in 2002 to permit Ethiopians in the diaspora with foreign citizenship to be treated as nationals, by offering a "Person of Ethiopian Origin" identification card (locally known as the Yellow Card) for foreign nationals of Ethiopian origin. The purpose of the new designation was to strengthen Ethiopian nationals' ties to their home country and also to facilitate their contributions to economic development by lifting existing legal restrictions and providing avenues for investment. The Yellow Card entitled the cardholder to most of the rights and privileges of an Ethiopian citizen such as entry into Ethiopia without a visa, the right to own residential property and the right to live and work in the country without additional permits.

Holders of the Yellow Card were accorded the same benefits and rights as domestic investors. These policies encouraged many in the diaspora to invest in small businesses in Ethiopia. Many investments at this level included cafes, restaurants, retail shops, and transport services in big cities and small towns that were once restricted to Ethiopian nationals living in the country. There is considerable variation in diaspora investment in Ethiopia by region and chartered city. The urban origins and orientation of most of the Ethiopian diaspora can be gauged from the fact that cities are their preferred areas of settlement in host countries, including the United States, and also their favoured locales for investment within Ethiopia. Even in the case of investments, the connections between the diaspora and the capital cities of Washington, D.C. and Addis Ababa are evident. Addis Ababa was the favourite investment destination, accounting for more than 90% of all diaspora investments in the country (Ethiopian Investment Agency, 2008). Greater perceived economic opportunities and security in the city and the relative ease of conducting business were incentives for investing in the capital city. Moreover, the diaspora had greater familiarity with the Addis Ababa, its institutions and networks, most having lived, studied or worked in it. An additional factor could be the Amhara-dominated diaspora's desire to continue to keep Addis Ababa as the dominant economic hub in the country even as the current Tigrinya-dominated government channels more resources to developing the northern province of Tigray.

Addis Ababa was also the locale of the first annual Ethiopian Diaspora Business Conference held in 2007 to explore investment opportunities and incentives for Ethiopian expatriates. Succeeding annual Diaspora Business Conferences were held in Washington, D.C in the United States in 2008, 2009 and 2010. At the conferences, which were attended by Ethiopians living in North America and even Western Europe, the Ethiopian diaspora was encouraged to invest in the home country and partner with Ethiopian producers to import products to their states and countries of residence.

Ethiopians living in metropolitan Washington have taken advantage of the special provisions for diaspora investors and initiated business ventures in Ethiopia. Many of these businesses were established in either Addis Ababa or the hometown of the immigrant entrepreneur, testaments to the power of translocal urban connections. A restaurateur who runs a popular eatery in Washington, D.C. used her savings to open a guest house in Addis Ababa, another Ethiopian immigrant who runs a travel agency opened a sister agency in Addis, while an engineer returned to the city to open his own garment manufacturing unit on its outskirts.

Globalization and the accompanying increase in mobility of persons and capital have not resulted in the disembedding of Ethiopian immigrant entrepreneurs. Rather, in the movement of immigrant capital and its investment, place matters. Especially if the primary home of the immigrant entrepreneur continues to be in the host country, there is a strong link between investment decisions and place-based belonging and personal ties as well as evidence of translocality in the choice of the site of investment. Although not usually accompanied by the permanent return of the diasporan investor, capital invested by the diaspora in Ethiopia also generates new urban forms, services and infrastructure such as gated communities, cyber cafes, shopping malls, travel agencies, schools, hotels and guest houses. The investor-entrepreneur's need for trusted and reliable persons to oversee and help in the construction of buildings, the establishment of the new business and its management has led them to seek local partners. Most diasporic entrepreneurs who only return periodically to the home country favour close family members or good friends as business partners and managers, another reason for establishing a business in the city or town from which the entrepreneur hailed.

A case in point is Bereket, who lived and worked in a number of cities in the United States for over two decades before he decided to start a business in Ethiopia. In 2003, he went to Addis Ababa with the intention of exploring the business environment and investment opportunities there. Motivated by encouraging feedback from compatriots and friends who had either returned permanently or had established new businesses in Ethiopia, as well as the financial and policy incentives put in place by the Ethiopian government for diasporan investors, Bereket decided to open a construction equipment rental business. However, he did not plan to relocate to Addis Ababa as his wife and children were reluctant to leave the United States and the Washington area. A compromize was reached when a friend whom Bereket had known since his high school days, who had lived in the same neighbourhood as him in Alexandria, Virginia and attended the same church, offered to be a co-investor and also manage business operations of the company in Addis Ababa. Today, Bereket visits Addis Ababa about twice a year, while his partner manages their company's operations on a day-to-day basis.

Despite, or perhaps, in response to residential dispersion in highly diverse neighbourhoods, in recent years Ethiopian immigrants in the Washington area have also made concerted efforts to maximize the group's political voice. Political activities of Ethiopian immigrants are not restricted to the jurisdictions in which they currently live or work. The Ethiopian American Council (EAC) is an organization that was established in the late 1990s by Ethiopian immigrants in the Washington area to create greater awareness in the United States about Ethiopia and its recurrent problems, particularly those related to famine. An added objective of the group at that time was to lobby the U.S. Congress to provide more aid to Ethiopia. Since then, the EAC has grown to become an ethnic lobbying group that has representatives and chapters all over the United States and a major objective of the organization is to effect changes in U.S- Ethiopia political relationships.

The political environment in the home country can be affected by and affect the diaspora especially if immigrants have strong political allegiances to certain groups or parties. Ethiopian immigrants (particularly those who are first-generation) closely follow political developments in the home country and shifts in political power. The political involvement of the diaspora in home and host country can be contentious. Most of the members of the Ethiopian community in Washington who left Ethiopia due to the socialist revolution of the 1970s are ethnic Amharas. As a group, they tend to be somewhat antagonistic towards the largely Tigrinya-dominated ruling party.

Many Ethiopian immigrants in the United States were distressed by the outcome of the May 2005 parliamentary elections in Ethiopia and the alleged suppression of protests that followed. Reportedly, thousands of demonstrators who protested the results of the 2005 elections that favoured Prime Minister Meles Zenawi and his party were detained, while over a hundred people died in election-associated violence. Although deemed peaceful, European Union officials have said that the elections did not meet international standards, while local protestors claimed that the elections were rigged. The EAC has rallied Ethiopians in the United States against Prime Minister Meles and pressed Congress to act through sanctions against alleged human rights violations, particularly the harsh treatment of political opposition in the country. A bill to this effect was introduced in 2006 and re-introduced as HR 2003, the Ethiopia Democracy and Accountability Act, in

2007. However, although the U.S. House of Representatives (Congress) passed the bill, it did not become law as it did not get to a Senate vote (Govtrack.us/2008).

M.P Smith (2008) notes that transnational actors are often connected via translocal ties to opportunities, structures and practices found in cities. Using translocal and transnational connections, the Ethiopian community in the United States and those in the Washington area in particular elevated what might be considered national issues in Ethiopia to those of international importance by framing the perceived problems with the 2005 elections and the current Ethiopian government in terms of universal human rights violations. Here as well, immigrant political activists have "jumped scale" by using a group that is anchored in Washington, D.C. but draws on the Ethiopian community from around the country to try to bring about policy changes that would affect the national governments and policies of both the United States and Ethiopia, asserting that the actions of the Ethiopian government were a threat to civil liberties.

Conclusion

It is evident that interactions among Ethiopian immigrants span different scales, identities, and localities. The multiple affiliations and simultaneous attachments to multiple places in the narratives of Ethiopian immigrants as seen in this chapter speak to complex forms of place-centered and place-based identities and actions for this group. Spread out over the United States, largely in its cities, translocal ties allowed Ethiopian immigrants to form and maintain a sense of belonging and stay connected. Although community has become increasingly portable, there continue to be geographic spaces in which ethnic gatherings take place, while the networks and activities of Ethiopian immigrants in the United States continue to be affected by relationships formed in the schools, colleges, churches and neighbourhoods in towns and cities of origin. Many of the encounters and linkages of the Ethiopian community bring together different local worlds and the points of reference for their far flung networks are grounded in localities. Significant among these places are the two cities that hold the greatest symbolic and material attachments for Ethiopian immigrants – Washington, D.C. and Addis Ababa.

The understanding of translocality as concurrent attachments to multiple places underscores immigrants' connections with and sentiments to both origin and destination settlement areas. Immigrants bring with them understandings of the places from which they hail, identifying themselves by a plethora of place-based labels. These linkages have implications for integration in the host country as immigrants use home country networks that operate both laterally and hierarchically to form community, obtain housing, jobs and various kinds of assistance, sometimes even before the actual move. With increasing length of stay in the host country, immigrants are likely to identify more with the neighbourhoods, localities and cities in which they reside and work, forming new configurations of place-based bonds. Hence, translocality becomes important in understanding the processes of mobility whether it be the movement from Ethiopia/Addis Ababa to the United States or secondary migration between U.S. cities as well as the outcomes of these processes.

Interconnectivity across and between scales complicates traditional notions of local, national and global. Instead of viewing the local as limited and circumscribed, or even of a local-to-global scale continuum, the concept of translocality provides a space where discourses, ideologies and actions between and across scales can be appreciated, comprehended and reconfigured. The strategies used by Ethiopian immigrants to accomplish immigrants' goals of greater empowerment through political organizing are at once local, national, translocal and transnational. Community negotiations of difference, inclusion and power are seen in the translocal networks that were formed to lobby for changes in US policy through the Ethiopia Democracy and Accountability Act of 2007 and its earlier versions, and the efforts to form a "Little Ethiopia". These efforts occurred at localized levels, but had city-wide, national and even international ramifications.

It is in through translocal networks, translocal processes and translocal places that the tensions between the global and local are often mediated. Networks are formed collectively by individuals and even though many immigrant networks are not place-restricted, they develop unevenly in space. Moreover, networks remain anchored in particular places, with cities (such as Addis Ababa and Washington, D.C.) often forming the nuclei where key actions related to the networks' mission occur and where principal actors and constituents are located. Translocal processes allowed Ethiopian immigrants to transgress spatial boundaries and renegotiate spatial hierarchies. Through vicarious imaginaries and actual mobilities, immigrants affect the perceptions and presentations of cities and city spaces; while through localized political and social movements, they sought to affect policies and actions at different scales.

Ethiopian immigrants and transnationals in DC and Addis Baba make deliberate use of translocality in order to advance their social, political and economic capital in both cities. This interpenetration of neighbourhood, city, regional and national scales and localities due to immigrant activities and activism complicates the notion of spatial hierarchies.

The immigrants' tactics reflect conscious strategies rather than mere coping mechanisms by a displaced people. Ethiopian immigrants built and used translocal affiliations and networks to their advantage, often transgressing hierarchical layers by jumping scale and bypassing intervening spatial levels of structure and authority. They used their political connections and political capital at the sub-city/neighbourhood level to try and make socio-cultural and political claims that would resonate nationally and even internationally through their mission to create a Little Ethiopia in Washington, D.C. They employed the strength of the Ethiopian population in the U.S. capital, the ties that were formed within the immigrant group and the ethnic solidarity of the Amharas as well as translocal ties to Addis Ababa to funnel investments to Ethiopia's capital city and place it at the forefront of urban development in Ethiopia. The multidimensional nature of translocal urban spaces in Washington, D.C. and Addis Ababa and the way in which these spaces can be multiply inhabited underscores the idea that localities are not static, but are continually being re-imagined, reconfigured and renegotiated through translocal processes.

PART 5:
Epilogue

PART 2

Chapter 11
Translocality: A Critical Reflection

Michael Peter Smith

In developing a research agenda for investigating transnational urbanism I tried to capture a sense of emergent social relations under globalizing conditions that are situated in specific places yet operate across geographical distance and are both embedded in and transgress processes of state power (Smith 2001, 2005a). I felt that this analytical optic usefully captured a sense of geographically distanced yet sociologically situated possibilities for constituting and reconstituting social relations. I viewed cities as key sites for emplacing communication circuits, organizational networks, economic linkages, and political projects that spanned national borders. Cities fostered a wide variety of such *translocal* linkages, I argued, because they concentrated the social, physical and human capital used to forge a multiplicity of socio-economic, cultural and political projects that linked localities across borders. For me, the best way to investigate transnational urbanism was to map these diverse translocal connections among the people, places and projects situated in cities at the crossroads of such ties. Put most simply, my take on transnationalism and the city focuses on the socio-spatial processes by which social actors and their networks forge translocal connections and create the linkages between and across places that I called *translocality*.

Situated yet Mobile Subjectivities

In approaching transnationalism from the vantage point of translocal connections I developed an understanding of translocality as a mode of multiple emplacement or situatedness both *here* and *there*. This differed significantly from the uses made of translocality by two leading social anthropologists, Arjun Appadurai and Ulf Hannerz, who understood translocalities as newly constituted socio-cultural spaces of mobility, displacement, and deterritorialization. They used the term to capture a sense of the fluid boundaries and identities thought to characterize a world of growing economic and cultural globalization.

In *Modernity at Large*, Arjun Appadurai used translocality both in opposition to and yet as a potentially constitutive element of locality, understood as the grounded neighbourhood sites of traditional ethnographic inquiry (1996: 192). For Appadurai translocalities differ from migrant ethnic enclaves but nonetheless influence the development of identities within them. They are imagined or virtual neighbourhoods that emerge at the articulation of media and mobility. As products of the increased global mobility of transnational migrants, Appadurai depicts translocalities as "deterritorialized imaginings" of ethno-national identity formation that implode into actual migrant ethnic enclaves, becoming agents in the production of a new sense of locality un-moored from the pull of the nation-state in which the real enclaves are located. Appadurai used terms such as "virtual neighbourhoods," "displaced public spheres," "delocalized transnations," "counter-hegemonic imagined worlds," and "translocal communities" interchangeably, investing these imaginings with the capacity to generate new "postnational identities" or "thoroughly diasporic collectivities" freed from "the linguistic imaginary of the nation state." (1996:166)

Appadurai's understanding of post-nationalism, like his construction of translocality, rests on the assumption that the power of nation-states to command loyalty and identity is declining in a world of highly mobile subject formation. He speaks of the waning power of receiving states to incorporate transnational migrants into loyalty to and civic engagement in their host society. In his view, "translocal communities" in the U.S. are described as being "doubly loyal to their nations of origin and thus ambivalent about their loyalties to America." (1996: 172) This challenge to the nation-state as an exclusive site of loyalty may also affect sending states in so far as the homogenous space of sending nations is also undercut by the global mobility of its former inhabitants, whose journeys bring them into contact with new places and capacities for imagining the self.

Appadurai's theoretical formulation of translocality focuses on the role of mass media and other modes of cultural globalization in shaping transnational migrant identities rather than their identities being enacted through their own actual socio-economic and political practices. As I have pointed out elsewhere (Smith 2005b), I was thus not surprised that the findings of my transnational ethnographic fieldwork on the social construction of new political spaces for citizenship across borders directly contradicted Appadurai's assertion of de-nationalization and a rising "post-national" consciousness across borders. In that body of work I examined the actual material practices of transnational migrants and other transnational actors (including state-centered actors and international development elites) in a historically specific institutional and political-economic context. One of the key findings of my research studies (Smith 2003, 2005b; Smith and Bakker 2005, 2008) was that successful civic engagement by trans-migrants in sub-national politics in their communities and regions of origin in Mexico generated a sense of political efficacy that prompted their active involvement in urban, regional and national politics in California. Their translocal interconnectivity led to a kind of borrowing and lending of everyday political experiences both here and there that led to institutional and policy changes on both sides of national political borders. Rather than uncovering *deterritorialized* identities and practices I found a *reterritorialization,* if you will, of political life on both sides of putatively self-contained borders.

Ulf Hannerz has been another influential voice conceptualizing translocality. Like Appadurai, Hannerz understands translocality largely in terms of heightened mobility rather than the situatedness of migrating subjects. In "Transnational Research" (1998: 239) Hannerz focuses the study of translocalities on sites that are central rather than peripheral "within the global order." He views peripheral sites largely as sedentary and un-dynamic spaces. In contrast, he views translocalities as sites that are "intensely involved in mobility and in the encounters of various kinds of mobile people..." Such spaces of encounter also tend to be "nodes" connecting transnational social and cultural processes. In discussing the ethnography of "translocality," Hannerz singles out as potential field sites such spaces as "hotels, airports and similar institutions" along with "world cities" and other "sites with a lot of transiency..." (1998: 239).

In "Power in Place/Places of Power: Contextualizing Transnational Research" (2005b) I have argued that this usage of translocality leaves little room for studying the social construction of less transient and more situated translocal connections such as the socio-cultural and political-economic ties being forged by migrant social networks linking small cities in California to even more peripheral villages in rural Mexico. I argued further that Hannerz's main formulation of translocality failed to consider that migrating subjects in one place may be linked to relatively sedentary *stayers* in another place by the practices of transnational households, village-based networks, or political projects and state policies seeking to situate the mobile subjects in such a translocal space. (Smith 2003, 2005b; Smith and Bakker 2008)

Upon rereading Hannerz's essay "Transnational Research," I recently discovered a more potentially promising use of *translocal* fieldwork found later in his essay. Commenting on the character of multi-sited fieldwork, Hannerz identifies what he describes as "relatively easy" cases of "multilocale" fieldwork, such as studying transnational migrants "at home and abroad" or investigating the practices of corporate elites "at headquarters and at branch sites." He then distinguishes these easy cases from more problematic instances in which, for example, multilocale research may also be "translocal." In this instance he defines the translocal as a *unit of analysis*, a "network of sites" where "parts of one's ethnography may have to be *between* these sites..." (1998: 247). Here Hannerz seems to mirror George Marcus's (1989) and my own (2001, 2005a) call for a multi-sited transnational ethnography that requires a precise investigation of, in Hannerz's words, "translocal linkages and interconnections between these and the localized social traffic" (1998: 247). I could not agree more with this formulation.

Unfortunately, however, Hannerz tends to undercut the point he is making about the need to study the everyday life-worlds enacted in these translocal social spaces by characterizing such spaces as "somehow deterritorialized." He thus worries that doing this sort of translocal ethnography would require a more *relational* view regarding the cohesion of the translocal units and the kinds of connections they embody, in contrast to the more traditional anthropological study of purely small-scale and local community formations. In positing this methodological conundrum in which "the local" is taken as the sine qua non of both *community cohesion* and *deep ethnography* Hannerz fails to take into account a mountain of evidence (e.g., the anthropological literature summarized in Gupta and Ferguson 1997 and Smith 2001) that the traditional local cultural formations that have formed the subject matter of conventional or deep ethnography have seldom been purely unitary cultural formations. Hannerz expresses fear that ethnographic thinness might characterize multi-sited translocal ethnographic research and assumes that this thinness can be replaced by qualitative depth simply by focusing the ethnographic gaze for long periods of time on "the local." Near the end of this essay, Hannerz does concede, however, that "insisting on carrying out an entirely local study in a site strongly marked by translocal and transnational connections would surely not result in satisfactorily complete, deep ethnography either." (Hannerz 1998: 248) This is a perceptive and most welcome concession.

In my view, social scientists should not approach the issue of translocal agency with theoretical preconceptions about the hyper-mobility and thus indecipherability of key sites and of the migrating subjects within them, as Hannerz appears to do. Nor should they be seduced by the new idioms of media generated post-national identity formation imagined by Appadurai. This is because in forming their own sense of agency the people one selects to interrogate as makers of and dwellers in translocal geographies are always already positioned or situated rather than unmoored, subjects. People "on the move" like more sedentary subjects, still occupy multiple, historically particular, social locations – they are raced, classed and gendered, with or without papers, possessing or lacking resources, tied to religious creeds or not, having differing degrees of political and social capital, and so forth. They act in historically specific contexts and are subject to the inner tensions and conflicts derived from their multi-positionality. Awareness of this multi-positionality of acting subjects and of the institutional, political-economic, and social structural contexts in which they act, forces us to think seriously about how mobile subjects are *emplaced*. Thinking about the sites of emplacement of mobile subjects helps guard against the macro-analytic view of mobility as occurring in an abstract, globalized "space of flows."

Ulf Hannerz is certainly correct when he speaks of the manageability of ethnographic research in simpler cases of multi-locale fieldwork such as studying transnational migrants "at home and abroad." Indeed, much of the literature on transnational political relations forged by migrant hometown associations stresses the depth and centrality of such bi-polar translocal social relations. Material and symbolic exchanges linking two localities across borders are shown to be key forms of translocal social capital promoting trust, sustaining collective solidarity, and generating community development projects across borders. The Mexican transmigrants studied by Robert Smith (1998) and Luin Goldring (1998), as well as my own extended case study of the translocality of Napa, California-El Timbinal, Mexico (M. P. Smith 2003; Smith and Bakker 2007), focused largely on translocality understood in this relatively straightforward sense. Bi-local relations and belongings generated a shared sense of interests and meanings that bound key actors in the translocal social field together, sustaining their sense of hometown membership and belonging. These social networks in migration and their attendant modes of social organization – economic remittances, social clubs, celebrations and other social processes based on translocality – constituted and sustained *translocal social structures*.

Differentiating the Translocal from the Transnational

Through my decade-long study of political transnationalism I have learned that when studying transnational connections it is often necessary to move beyond such dyadic local-local connections that constitute translocalities, and which, when I wrote *Transnational Urbanism* (2001) I had viewed as a necessary starting point, if not the most reliable basis, for researching transnational connections. In *Citizenship across Borders* (2008), a book that explores various dimensions of US-Mexican political transnationalism, my collaborator Matt Bakker and I were required to consider in detail both the translocal and the transnational practices of our interview subjects and *their relationship to each other* as well as to take into account the wider institutional, socio-cultural, and political-economic contexts in which the politics of translocal community development and other forms of transnational political practice such as transnational electoral politics and transnational political reform movements were embedded. These included studying, in a wide array of locations, the beliefs and practices of transnational migrants, as well as the ideologies and practices of state-centred actors in Mexico and the US, and the activities of a myriad of non-state actors such as international development elites and the leaders of non-governmental organizations, all of whom sought to shape the character of the translocal, transregional and transnational politics we were studying.

Through these qualitative research experiences, conducted in multiple locations and operating at multiple scales, I have reached the conclusion that just as not all translocal connections are necessarily transnational, not all transnational connections are necessarily translocal. Because of the complexity, if not the messiness, of my encounters with more complicated crisscrossing political geographies, I have come to fully agree with the multi-scalar epistemological view recently voiced by Peggy Levitt (2004: 3) namely: "It is critical to examine how [transnational] connections are integrated into vertical and horizontal systems of connection that cross borders. Rather than privileging one level [e.g. the local] over another, a transnational perspective holds these sites equally and simultaneously in conversation with each other and tries to grapple with the tension between them."

To take one example of this multi-scalar conversation, consider my study of the transnational electoral politics of the "Tomato King." In seeking office as municipal president of his hometown of Jerez, Zacatecas, Andrés Bermúdez, who came to be known as the "Tomato King," did not rely on political support from translocal networks of circular migrants between Jerez and his place of residence and successful agriculturally-related business operations in Winters, a locality in Northern California. Bermúdez's political ties to Mexican transnational migrants were not simply local to local but spatially very dispersed, encompassing Zacatecan migrant federation leaders in Los Angeles and Southern California as well as potential campaign donors and return migrant leaders currently living in other U.S. states and coming from different Mexican states.

Nor did Bermúdez discursively frame his candidacy as a quest to establish a "Jerezano" translocal identity. Rather, the Tomato King's political strategy, discourse, and appeal were universalistic, multi-stranded, and highly mediated by his global media coverage and his wider transnational political connections. Investigating this multi-scalar reality required us to travel to a much wider array of places in California and in Mexico than did my earlier study of the making of the translocality of Napa-El Timbinal. To make sense of how a transnational electoral coalition connected diverse spaces and networks of power on both sides of the US-Mexican border, the number of places, the types of actors interviewed, and the forms of transnational interconnectivity had to be significantly expanded beyond local-local ties.

Moving beyond the practices of bi-local migrant networks, Bakker and I supplemented our ethnographies of key migrant leaders and the people, places, and organizations they were tied to by conducting extensive in-depth qualitative interviews with state level political elites and their policies in sending and receiving states, regions, and localities; non-migrating actors like social movement activists, public intellectuals, and members of the existing political class in Mexican state and local politics; and interested political observers from other Mexican and US cities and states, such as Houston, Texas and Chicago, Illinois, who sought incorporation into the political network of networks being forged by this transnational coalition. To study this crisscrossing, multi-scalar political process we necessarily paid close attention to the changing geographical connections among and the changing culture and politics of the specific neighborhoods, cities, and regions that were the grounded sites linked together by these migrant leaders' political practices. The complex politics underlying the formation of this US-Mexican transnational coalition thus moved well beyond the bi-local scale. Indeed, it resembled the scale-jumping, hetero-local, multi-scalar politics of the Ethiopian diaspora in Washington, D.C. insightfully discussed elsewhere in this volume by Elizabeth Chacko.

Beyond specific neighbourhoods, cities, and regions, Bakker and I also carefully attended to the ever wider political-economic, geo-political, and institutional contexts in which migrant leaders' translocal and transnational practices were taking place and in which the mobile subjects we were studying were emplaced. We found this move important methodologically because it guarded against a de-contextual ethnographic inscription of transnational *or* translocal "communities" as timeless cultural wholes detached from the often-contested historical and geographical contexts of their emergence. Such romantic social constructions of community formation tend to obscure the ongoing power relations underpinning the formation and reproduction of any kind of "community," transnational, translocal or otherwise.

To sum up, the mobile subjects I have studied may be globally mobile *movers* but they nonetheless remain situated within various power-knowledge venues and occupy classed, gendered and racialized bodies in space. As transnational migrants they move to and live in real rather than virtual neighbourhoods, cities, and regions and they use both their imaginative and their material resources to try to improve living conditions in both the real places they came from and those in which they currently live. Their practices underline their social connectedness and multiple emplacement across borders rather than their emancipation from assimilative social structures.

The Multi-scalar Imagination

It is more essential than ever to bring a multi-scalar imagination to translocality research in general and to future research on the politics of translocal place making in particular. This is one of the chief virtues of *Translocal Geographies*. Taken together, the contributors to this book have brought a fruitful multi-scalar imagination to the dialectic of mobility and place making. Some of these studies move our imagination from the metropolitan scale of translocal cities to a variety of *smaller* micro-structural scales of material life and social practice including residential neighbourhoods of migrant passage within London; elite hotel leisure spaces in Buenos Aires and the opportunity structures they foster both within and across national borders; a sacralized urban tourist space in Cambodia and the ties there binding tourist workers to their Cambodian ancestral villages; and even domestic home-making spaces connecting highly skilled returning British migrants from Singapore to both ends of their translocality, in this instance a transnational one.

Other studies in this book, particularly, those by Ben Page, Katherine Brickell, Elizabeth Chacko, and Ryan Centner parallel my own work (Smith and Bakker 2005; Smith 2007; Smith and Bakker 2008) by demonstrating that it is often necessary and also fruitful to take into account *wider* scales or macro-structural settings when investigating translocal geographies. Such scales include, but are not limited to, sub-national regions and the social, political, and imaginative ties they foster; nation-states and the policies they pursue to channel or otherwise adapt to the transnational mobility of their citizens; the global investment policies of states that facilitate the flow of diasporic financial investment to specific cities; trans-state relations, policies, and institutions and their historical impacts on migration flows and global mobility; and the differential socio-spatial, political-economic, and cultural impacts of the neo-liberal regime of global governance and the opportunities and constraints it engenders for the pace and character of global mobility.

In what ways is translocality understood in the multi-scalar studies included in *Translocal Geographies?* What spaces and places are taken to constitute the terrain of translocal geographies? How are they similar to or different from the translocal ties that have become a prominent feature of research on transnationalism? When I asked myself these questions the first thing I noticed is that several of the local-local connections studied in this book take place *within* rather than across national frontiers. This should not be surprising since there always have been vast numbers of people involved in internal migrations. Moreover, while there has been historic growth in the absolute size of mass migration and spatial mobility in recent decades (Urry 2007; Castles and Miller 2009) a good deal of this mobility has occurred within rather than across national borders. Indeed, as I have pointed out elsewhere, economically driven rural to urban mobility within nation-states has long been recognized as a driving force of world urbanization and was identified by the Chicago School of sociology over a century ago as a constitutive indicator of modernity (Smith and Guarnizo 2009). This dynamic, along with forced internal migration within war-torn countries in at least three continents, has engendered all sorts of translocal ties ranging from fragmented households seeking to reproduce themselves in more than one location to the production of refugee camps filled with people seeking a right to return to their places of origin. Just as I have come to conclude that not all transnational connections are translocal, these realities, along with many of the richly textured translocal case studies found in this book, clearly support the proposition that not all translocal connections are transnational. I thus concur with a key premise of this volume that scales other than the transnational are important sites for negotiating the social, political, and cultural aspects of translocal migration and mobility.

Beyond this, I also am in basic agreement with the conceptualization of translocality deployed by Katherine Brickell and Ayona Datta, the editors of *Translocal Geographies*, in framing this book. Brickell and Datta reject the representation of translocalities as purely imagined communities or as globalized spaces of hyper-mobile flows. Rather, they define translocalities are interconnected spaces of "locatedness" spanning multiple sites of material life both within and across borders. They treat migrating subjects as irrevocably situated, moving across multiple spaces that re-locate them within shifting power-knowledge venues, against which, and sometimes through which, they act to shape the conditions of their own mobility and existence. Following from this theoretical move, *place* is fruitfully re-imagined as a social space standing in opposition to what Tim Cresswell (2006) has called a "sedentarist metaphysics." The places constituting the translocal geographies in this book are conceived as dynamic yet situated sites of becoming; as unique articulations of situatedness through which migrating subjects must act as they seek to orchestrate lives both within and across national frontiers.

The virtue of this move is that it transcends sedentarism while avoiding the equally confining metaphysics of nomadism found in some cultural studies accounts of global mobility and hybrid subjectivity (e.g., Deleuze and Gauttari

1986, 1987; See also Katharyne Mitchell, 1997, for an excellent critique of "hybridity" in transnational studies). For me, romanticizing "nomadism" as a deterritorialized way of life existing outside the organized state is reminiscent of Hannerz's celebration of "people on the move" and the hyper-mobility found in international airports and other sites of transiency. When understood from the standpoint of people negotiating translocal connections, I think it is more fruitful to think about "dwelling in motion" as a means not an end, and usually as a means to a socio-cultural, economic, or political project envisioning a more settled mode of human existence in the face of the unsettlement of dwelling-in motion whether for self, family, village, region, nation, or trans-nation.

I am not here arguing against the study of the social life of nomads or any other kind of people on the move, but rather against weakening the notion of *situated subjectivity* and *emplacement* as key elements in the making of translocal connections. When the semantics we appropriate to represent human mobility are too fleeting, ephemeral, and unbounded, we move from a world where social structures still matter to a world of pure flexibility, deterritorialization, and disembeddedness. Celebratory discourses of hyper-mobility and nomadism fail to take into account the structures of domination and venues of power/knowledge that still facilitate or constrain human mobility and connectivity in the word in which we live.

Social Constructions of Home, Belonging, and Return

In her reflective essay on translocal geographies of return migration (this volume), Madeleine Hatfield (née Dobson) discusses the return of professional middle class economic migrants to various British localities from work experiences in Singapore. Her chapter envisages translocality as operating at the hyper-local scale of specific households and the imaginative resources and material objects emplaced in their homes. She depicts domestic home spaces in the UK as enduring locales for migrants, even when they are physically absent living abroad. While abroad, these skilled migrants symbolically and imaginatively incorporate elements of the homes they have left behind into the homes they make in Singapore. A useful method employed in this research is Hatfield's focus on following the "thing," namely aspects of material culture, to discern translocal connections that might otherwise be taken for granted or difficult to articulate. In the UK, such things include the display of souvenirs from abroad, such as Buddha statues, which, according to Hatfield, turn participants' UK homes into translocal sites. In Singapore, another object of emotional connectivity is a photograph of a house in the UK, frozen in time, that presences migrants' distant home while they continue to live in Singapore. From this perspective, "coming home" for return migrants is less a return to a national level homeland or the recovery of what Anthony Giddens (1990: 19) would call the distantiated relations that have penetrated their current "locale" in a specific site of return (although migrant "return" also may operate in these ways). Rather, it is a return to the symbolism and materiality of "domestic home spaces" – specific houses, pieces of land, loved ones, or cherished spaces and places of previously transnational families.

By my reading, this sort of iconic rendering of past experiences may, as is argued by Hatfield, constitute markers of translocal connectivity. But they also may be interpreted as efforts to "freeze the past." As such, they tend to represent a nostalgic longing to "belong somewhere else" – in a dwelling, in a neighbourhood, in a city, in a different nation. Ironically, such iconic markings of the social spaces of one's past, whether here or there – actually may prevent one from achieving an openness to culturally different experiences and the acquisition of multiple cultural repertoires that are often regarded as key attributes of both transnational and translocal belonging – i.e., the formation of a sense of belonging *both* here and there (on this double consciousness see Smith 2007). One is reminded by Hatfield's narrative less of the self-affirming social construction of translocality as an emergent duality than of the more self-limiting material practices of the "accidental tourist."

Of course, nostalgic attachments are not necessarily self-limiting. In fact, Brian Tan and Brenda Yeoh, in their chapter on the reproduction of rural-urban translocal family relations amongst the Lahu in Thailand (this volume), provide an eloquent defence of nostalgia as a means to both conserve memories of locality and to imaginatively help reproduce translocal family relations. Tan and Yeoh point out that nostalgia, understood as an attachment to the past that helps make sense of people's present lives, plays a major role in the conservation of a sense of locality in a "left-behind village" in Thailand. They treat the "left-behind" village of the rural Lahu as a translocality in the making, tied to various social and economic networks, including translocal family networks formed as increasing numbers of the children of the parent generation migrate out of the village. In this instance, nostalgic memories embodied in physical objects are used to provide stability to parental lives in the village community against the effects of translocal influences and to maintain attachments to their children living in more cosmopolitan urban locales. One mode of agency of "left-behind" family members is a parental strategy of conserving their children's personal belongings and dwelling places to give the parents a sense of stability in the present and hope for the future of their translocal families and their changing local village in the face of new pressures emanating from the wider array of translocal linkages now effecting the village.

Related questions are addressed in Katherine Brickell's study "Translocal Geographies of 'Home' in Siem Reap, Cambodia" (this volume). Siem Reap is the home of the UNESCO-designated tourist and heritage site of Angkor. Brickell's ethnographic study focuses on inhabitants' evolving relationships to "home" as a metaphorical, and potentially multiply-sited space of personal attachment, belonging, and identification. She tells the story of how rural Cambodians who have migrated to Siem Reap to take advantage of new employment opportunities in tourism in the face of diminishing opportunities in traditional rural farming construct a sense of belonging. These internal migrants are currently physically emplaced in Siem Reap but remain connected imaginatively to rural people and places, forming a vivid memory of their ancestral villages of origin. They are also tied materially to people in these other localities by their use of translocal networks to transmit money, things, mobile-phone messages of love and support, and other ties that bind them to people in these other places.

Yet Brickell's research subjects' sense of belonging is not only tied to rural ancestral homelands. Indeed, because they are living participants in the political economy of global tourism in a world tourist site that celebrates their national heritage, these urban migrants are also imaginatively linked to the discursive domain of the global and the national. Their "geographies of home" are also simultaneously characterized by desires to participate in what they regard as Cambodia's global development, as spearheaded by Angkor. They perceive Siem Reap as a key pole of tourism facilitating Cambodia's entrance into the world marketplace, thus enabling inhabitants to obtain useful material and social capital. They thus value both their mobility and their ancestral heritage. They locate themselves in different geographies simultaneously, using their imaginative and material resources, to feel at home with the local-local ties they have constructed yet also to maintain a sense of belong to a nation and fitting into the global cultural and political economy.

Perhaps the most unusual form of migration, belonging, and return analyzed in this volume is found in Christou's interesting analysis of the counter-diasporic "return" migration of second-generation Greek migrants from the world cities of Berlin and New York to enact translocal everyday lives in Athens, a city of "origins" which they have constantly imagined, but in which they have never previously lived. What sense of place, homeland, displacement, and otherness does this particular cohort of the Greek diaspora experience and enact as its members "return" to an Athens they have only ever known imaginatively? Based on the narrated life-stories of 60 Greek-American and Greek-German "returnees" the author paints a picture of complexly negotiated identity formation that entails a renunciation of the migration narratives of family betterment that motivated their parents to migrate from Greece in the first place. This is accompanied by a practical "leaving behind" of the migrants' nuclear and extended families that remain in the US or Germany. The move, in turn, is reinforced by their development of a new migration narrative, which the author calls a "personal myth." This myth is enunciated as a personal plan of action in which relocation to the ancestral homeland will offer a life of new beginnings, despite often-extensive empirical evidence to the contrary.

These second-generation "relocatees" (more accurate than returnees) experience an often jarring disconnect between their imagined homeland and the actual experience of everyday life in Athens – a world of recurrent institutional disorganization, exclusion, and xenophobic rejection of their journey of "return." The migrants nonetheless tend to stay in Greece and strive to realize their chosen plan of action. While no longer romanticizing their chosen city or nation as their place or their home, their personal narrative, as an expression of their agency, continues to be lived out. Christou does not explain why this personal myth of new beginnings motivates the social actions of her subjects. An explanation drawn from the voices of the migrants themselves would seem to be a useful next step in translocal research of this sort, if only to prevent readers from speculatively filling in the blanks with ungrounded assumptions about psychological denial or sociological intergenerational escape.

However we choose to address the conundrums of translocal identity production, cultural reproduction, or self-actualization in the face of increased global mobility, these analyses of the multi-stranded geographies of "coming home" remind us that all research on memory, belonging, imagined homes, and return necessarily raises basic questions of power. Reflecting on the dynamics of "return" in particular requires researchers of translocal connections to raise questions about wider scales of power and social practice. For example: Who has the power to return, however defined, and who does not? Which social classes and gendered bodies can most readily go home? Which migration statuses allow and which preclude return, or at least render it problematic (e.g., the status of political refugees). In what ways have the cherished sites of return themselves been exposed to significant cultural and structural transformations? To what home, therefore, is one returning? Why is the myth of return so often applied to low-income transnational migrants and refugees from the global South to North, who long for return to their necessarily changing homelands, but less often to less disadvantaged migrants returning from South to North, North to North, or within nation-states from cities to ancestral villages?

Translocality and Social Mobility

These reflective questions bring to mind another common theme informing research on migration, mobility and translocality, namely, the relationship between geographical and social mobility. While not explicitly framed in this way, two other contributions to this book, Ryan Centner's study of the uses of an elite hotel and a discotheque, viewed as networked translocal spaces in Buenos Aires, and Ayona Datta's study of the residential mobility of Polish migrants in London, can be interrogated for the light they shed on this theme. In Centner's "Ways out of Crisis in Buenos Aires," a desire for upward social mobility is a key driving force underlying the social construction of translocal networks forged by some native inhabitants of Buenos Aires, to help them weather the turbulence of global economic crisis. Two small social spaces within a redeveloped waterfront neighbourhood – an elite hotel and a discotheque – are represented as "translocal landscapes," useful sites where some users of these spaces have been able to activate "mobile resources" to improve their life chances.

The elite hotel is shown to be a translocal space where some hotel workers have managed to mobilize their own limited resources, e.g. – "just enough of an accumulated education, a European grandparent, or a knack for customer service"

– and combine them with the social, economic, and cultural connections possessed by guests in the hotel from cities in the US, Europe, or Australia to promote their spatial relocation and upward social mobility. Transnational spatial mobility in the global political and cultural economy is sought because it enables these networkers to become re-emplaced in less turbulent urban spaces across borders.

Social mobility within as well as outside Argentina is enacted, according to Centner, by some of those networking in a second small scale translocal venue – a discotheque in the redeveloped port area– where more promising occupational or residential "elsewheres" both at home and overseas are sought out and translocal connections to these places are forged as a kind of hedge against peoples' current life chances in Buenos Aires.

Simply being in these two small-scale sites however is no guarantee of either spatial or social mobility. As Centner concludes, the port-side neighbourhood and its privileged micro spaces "constitute a translocal landscape that can be activated by certain urban social groups with the know-how for directing their emplaced practices beyond Buenos Aires, while it remains inert and essentially resourceless for others who cannot effectively exploit translocal connections." Reflecting on this well crafted study, and its implications for future research on the production of translocality, I wondered what sorts of resources, aside from their possession of social and cultural capital, might distinguish successful users of "mobile resources" from the many other social actors situated in translocal landscapes who remain mired in their current urban fortunes.

In "The Translocal City," Ayona Datta also links spatial mobility to social mobility. Datta investigates both the transnational mobility of Polish migrants and their translocal movement across different neighbourhoods within London. At the transnational scale, the connection between the migrant workers spatial and social mobility is clearest. For instance, when one of Datta's interview subjects first moved between Poland and the UK he was without papers and thus held the status of an "illegal." He was forced to work for several years at sub-minimum wages, encountered non-paying clients, and did not return to Poland. After he acquired EU citizenship in 2004 he quickly moved up the ladder of social mobility, owned his own contracting company and frequently returned to Poland. His work, residential status, and opportunities for further upward social mobility were fundamentally restructured by this changed legal status. As Datta observes, the post-2004 identities of her interview subjects "are increasingly shaped through the regional space of the European Union, which is as much political, geographical, and structural – guiding the possibilities of 'free movement', work and dwelling across 'European' space."

The lion's share of Datta's chapter deals within Polish migrants' translocal mobility across different social spaces within metropolitan London. She shows how different London neighbourhoods become sites of local-local attachment as well as trajectories of migration and mobility among her interview subjects, enabling them to retrieve a materially rooted sense of agency and belonging. Initially prompted by a desire to live near their places of employment in the construction trades, and to minimize the cost of their housing and the time and costs of their journeys to work, these Polish migrants moved frequently from place to place, choosing to stay in temporary accommodations. To overcome a sense of transiency that this might entail, they sought to make these temporary sites into places "where they could mobilize a sense of belonging, familiarity and comfort." In short, they enacted in these sites various place-making practices that put them in contact and sometimes conflict with other users of these spaces, including other transnational migrants. The cheap neighbourhoods where Polish construction workers were able to afford shared accommodation also housed substantial numbers of other minorities from Asia, the Middle-East, and the Caribbean, with and against whom they have had to engage in negotiating spatial and social mobility while gaining a sense of emplacement and belonging.

Viewing the city as the quintessential site of these everyday translocal experiences, Datta argues that these encounters with others are key elements in migration narratives of social mobility, as migrants seek to improve their everyday lives economically, spatially, and socially. Clearly, the migration of Polish construction workers to London and its neighbourhoods is not an unproblematic journey of frictionless upward social mobility. Disappointment and disillusionment with their disconnection from the city centre produce lost dreams of living modern lifestyles at the core of a key site of Western capitalism. But the everyday experiences of the Polish migrants may also produce a sense of familiarity with the multi-ethnic residential neighbourhoods of East London that they share with other migrants, and comfort in those neighbourhoods that are also home to a network of co-ethnic friends and acquaintances to add to the ethnic food, affordable restaurants, and cheap rents they enjoy there.

In seeking to situate her mobile subjects, Datta usefully describes the inter-ethnic engagements, tensions, and accommodations involved when these ethnographic subjects encounter urban micro-spaces occupied by ethno-racial "others." She subtly describes the Polish migrants' engaged yet incomplete efforts "to understand, negotiate, and rework difference within particular spaces and places of everyday life in London," ranging from pubs and nightclubs to building sites and Asian takeaway restaurants. Datta characterizes these encounters as "everyday situated cosmopolitanisms," through which the Polish migrants' various ethno-national constructions of difference give a different, more fractured and differentiated socio-spatial meaning to the city of London than that of the unitary global metropole of which they had dreamed.

The Politics of Translocal Place-Making

The final contributions to *Translocal Geographies* explicitly address the contested politics of translocality. The politically contested space featured in Wise's "You wouldn't know what's in there would you?" is the main shopping street of Ashfield, a suburb of Sydney, Australia. In the past decade this once multi-ethnic shopping area, that had contained a mix of Anglo, Italian, and Greek shops, has been transformed into a "Little Shanghai" in which 85 percent of the businesses on Ashfield's high street are Chinese small business, mainly restaurants and small food an ethnic commodity markets. The streetscape is now replete with Chinese language signs and the signage has become a hotly contested political issue that has alienated other residents, engendered a highly emotional tabloid debate, and led to discussions at local planning meetings about the possible redesign of the signage.

For the new transnational Chinese migrants the signs, as well as the restaurants, food, music, and video shops featuring familiar ethnic wares, constitute what Wise nicely characterizes as "intensive diasporic practices of place-making" that recreate spaces of "embodied belonging" linked symbolically and materially to transnational spaces of origin, creating a vibrant translocal sense of "home away from home." But these modes of translocal belonging among the Chinese residents have also constituted a sense of local displacement among many non-Chinese residents who remained in the neighbourhood after its transformation.

Newly translocalized neighbourhoods are not empty meeting grounds. To those who maintain attachments to the old urban landscape, this familiar landscape has been radically erased, replaced by unfamiliar shops with signs deemed unreadable and unwelcoming. Wise hopes that over time a politics of "multicultural, translocal place-sharing" can be configured to negotiate these cultural and political conflicts over urban place making. While hope may spring eternal, the evidence derived from a similar neighbourhood multi-ethnic conflict over the transformation of a shopping area and its Chinese signage in Monterey Park in suburban Los Angeles (Horton 1995; Saito 1998) suggests that in the short run, at least, the conflict will endure and accommodation will be difficult.

A more complex and multi-stranded narrative of the politics of translocal place making is found in "Translocality in Washington, D.C. and Addis Ababa," Elizabeth Chacko's stimulating contribution to this volume. Chacko's chapter carefully specifies the translocal social spaces being forged and transnational financial linkages being constructed by the Ethiopian diaspora in these two capital cities. Chacko uses "translocality" as a lens for grounding the processes of international migration from Ethiopia to the U. S., pinpointing the secondary migration of Ethiopians across U.S. localities, detailing the local and transnational politics involved in the yet to be realized diasporic effort to construct a "Little Ethiopia" neighbourhood space in Washington, D.C., and structuring the flows of financial capital from the Ethiopian diaspora in the U. S. to various urban spaces in Addis Ababa.

While focusing on the grounded cities and neighbourhood spaces being linked together in translocal geographies, the multi-scalar imaginary clearly informs Chacko's thinking. She precisely captures the politics of "scale jumping" by which the Ethiopian diaspora in Washington, D.C. forged a relationship with a U.S. local neighbourhood politician whose influence was citywide and subsequently took him to Addis Ababa to introduce him to local politicians there, in order, she states, "to solidify their political connections and strengthen their platform at multiple scales." One of the main objectives of their political platform in Ethiopia has been to forge a fruitful political climate for diasporic economic investment in homeland urban spaces, which in turn, have successfully created new economic landscapes, services, and infrastructure in Addis Ababa ranging from gated-communities, to hotels, shopping malls, and even schools. The politics of place making thus operates at many scales and weaves together complex modalities of interconnectivity across and between scales that complicates traditional "nested" notions of local, national and global power and space. As Chacko rightly concludes, the concept of translocality deployed in her work provides a means whereby space transforming "discourses, ideologies and actions between and across scales can be appreciated, comprehended and reconfigured."

Because it is multi-scalar, the politics of place making necessarily crosses many boundaries. The meaning-making practices entailed in traversing borders and boundaries are well discussed in Ben Page's chapter "Fear of Small Distances: Home Associations in Douala, Dar es Salaam, and London." Page's study narrates the historical and contemporary practices of different migrant hometown associations with close affiliations operating within the nation as well as transnationally. These migrant clubs, some existing for decades, suture together often-passionate place-based loyalties, linking various localities within Tanzania and Cameroon to their core cities of Dar es Salaam and Douala or to London or Dallas. Page's study of these hometown clubs reveals that the migrants' diasporic practices, whether translocal or transnational, actually reflect an ideology of "localism," that produces fervent "urban loyalties" that sometimes trump loyalties to nation-states.

Page uses archival evidence to trace the development of hometown associations in these two African nations to the early 20th century. With respect to politics of translocal place making, Page observes that the migrant members of these hometown clubs were translocal in the 1920s, 1950s and even the 1980s without ever having left their nation of origin. Paradoxically, during the period of nation-building, the various hometown associations within nations often competed with each other for recognition and support and this inter-ethnic competition was often sponsored by national governments whose elites, in their own quest for political support, practiced a kind of clientelism that tended to favour "one community over another at one moment" and then reversed course at another time, redirecting their patron-client munificence elsewhere.

Interestingly, unlike hometown associations in other parts of the world, which often focus on the community and economic development of their hometowns, the main activities of the associations studied by Page appear to focus on improving the well being of club members in the urban milieus in which they are currently living, be it Dar es Salaam or the Dalston neighbourhood of Greater London, rather than their sending villages of origin. Page found that amongst migrant interviewees currently living in the UK, particularly from Cameroon, it was not uncommon to find people "who had never physically visited the town that was apparently the central justification of the association of which they were a member in the UK." Their identities were more closely tied to the major city back home that was a key site of memories of childhood, family, and friendships. Thus the key poles of their translocal identity were more likely to be urban-urban ties constructing a translocality of Douala-London or Dar es Salaam-London than the more typical rural-urban ties connecting current metropolitan locations to sending villages.

Page's painstaking investigation of the politics of place-making in city and nation in Cameroon, in particular, reveals that to preserve its power following the return of multiparty politics to that nation after a twenty year hiatus, the ruling party took advantage of extant ethnic competition, orchestrated, in part, by regional hometown associations, to prevent opposition parties from developing national unity or a nationally coherent ideological programme. As a result, he concludes, there has been a rescaling of the city relative to the nation in Cameroon that differentiates it from Tanzania. More generally, he concludes that "urban patriotism," defined as a deep-seated attachment to a particular city or cities, may be becoming more important at the current moment than loyalty to the nation-state. Page's fascinating analysis of the urban political identities produced by the politics of home-town associations in Cameroon and Tanzania, invites comparison with the politics of place identity involved in the operation of hometown associations in other world regions, such as Latin America, which have focused largely on the social construction of attachments to remote rural sending villages and their surrounding regions rather than to trans-locally connected urban spaces. Put more broadly, such comparisons suggest that the rescaling of place attachments may well become an important research theme in future translocality studies.

Conclusion

If we define politics broadly, the politics of translocal place making encompasses not just these last two contributions but also this entire book. I am thinking here of the micro-politics of household reproduction and dissolution in several of the case studies; the development of tourist spaces symbolizing global luxury consumption or national cultural heritage; the salvaging of left-behind villages; the everyday encounters of second-generation "returnees" with the xenophobia of the native population; the use of migration to claim a right to transnational labour mobility; and the negotiation of everyday encounters in multi-ethnic neighbourhood spaces.

Beyond these specific dimensions of translocal place making, the contributions to this book have helped me to think about several more general research questions that can be expected to shape future research directions on the making of translocal spaces. These include:

- What is the character of the macro-structural and micro-structural venues of power-knowledge that facilitate or constrain human mobility across translocal spaces?
- How do these differ when the translocal spaces in question are constructed and negotiated within rather than across national borders?
- How, precisely, does the multiple emplacement of situated subjects affect their sense of self and other, their access to material and symbolic resources, and their capacities to pursue their chosen life projects?
- What are the causes and effects of mobile subjects' social constructions of home and "away," belonging and exclusion, departure and return?
- What role does nostalgic attachment to people, places, and things play in the social construction of translocal projects?
- What key roles do networks of all kinds – economic, social, political, familial, locality-based, and technological – play in the production or transformation of specific translocal geographies?
- What roles do historically specific political logics and state policies play in the production or transformation of translocal spaces?
- Perhaps most importantly, in studying translocal geographies, how can we best cultivate a keen awareness of the role of imagination in social life and the corresponding responsibility of researchers to recognize the need to deploy a multi-scalar imagination in social research?

In the last instance, it is hard to avoid the conclusion that the making of translocal geographies, like the making of all geographical formations, involves the social construction of space as place. As such, translocal geographies are necessarily political constructs. "Making place" translocally is a meaning-making practice. It involves unavoidable questions of power – the power to name and claim space, knowledge of the stakes involved, awareness of changing opportunity structures, and the capacity to contest the practices of others. This is what makes the study of translocal geographies so intriguing.

Bibliography

Aarons, D. 2004. Traveling to Ethiopia For Clues About D.C.: Graham to Study Immigrants' Roots. *The Washington Post*, August 5, page DZ03.

Acharya, S. 2003. Migration Patterns in Cambodia – Causes and Consequences. Ad Hoc expert group meeting on Migration and Development, 27-29 August, Bangkok.

Aguettant, J.L. 1996. Impact of Population Registration on Hill Tribe Development in Thailand. *Asia-Pacific Population Journal*, 11(4), 47-72.

Ahmed, S. 1999. Home and Away, Narratives of Migration and Estrangement, *International Journal of Cultural Studies*, 2, 329-347.

Ahmed, S.C., C. Casteneda, A. Fortier and M. Sheller. eds. 2003. *Uprootings / Regroundings,Questions of Home and Migration*. New York: Berg Publishers.

Al-Ali, N. and K. Koser. eds. 2002. *New Approaches to Migration? Transnational Communities and the Transformation of Home*. London: Routledge.

Allen, J. and E. Turner. 1996. Spatial Patterns of Immigrant Assimilation. *Professional Geographer* 48(2), 140-153.

Al-Sayyad, N. and A. Roy. 2006. Medieval Modernity: On Citizenship and Urbanism in a Global Era, *Space and Polity*, 10(1), 1-20.

Aldous, J. and D.M. Klein. 1991. Sentiments and Services: Models of Intergenerational Relationships in Mid-Life. *Journal of Marriage and the Family*, 53, 595-608.

Amin, A. 2002. Ethnicity and the Multicultural City: Living with Diversity. *Environment & Planning A*, 34, 959-980.

Amin, A. 2008. Collective Culture and Urban Public Space. *City*, 1:2, 5-24.

Amin, A. and N. Thrift. 2002. *Cities: Reimagining the Urban*, Oxford: Polity.

Amit-Talai, V. 1998. Risky Hiatuses and the Limits of Social Imagination: Expatriacy in the Cayman Islands, in *Migrants of Identity: Perceptions of Home in a World of Movement*, edited by N. Rapport and A. Dawson. Oxford: Berg, 41-59.

Anguita, E. 2003. *Cartoneros: Recuperadores de Desechos y Causas Perdidas*. Buenos Aires: Norma.

Appadurai, A. 1990. Disjuncture and Difference in the Global Cultural Economy, *Public Culture*, 2(2), 1-24.

Appadurai, A. 1991. Global Ethnoscapes: Notes and Queries for a Transnational Anthropology, in *Recapturing Anthropology: Working in the Present*, edited by R.G. Fox. Santa Fe, CA: School of American Research Press, 191-210.

Appadurai, A. 1993. Patriotism and its Future, *Public Culture*, 5(3), 411-429.

Appadurai, A. 1995. The Production of Locality, in *Counterworks: Managing the Diversity of Knowledge*, edited by R. Fardon. London: Routledge, 208-229.

Appadurai, A. 1996a. *Modernity at Large: Cultural Dimensions of Globalization.* Minneapolis and London: University of Minnesota Press.

Appadurai, A. 1996b. Sovereignty without Territoriality: Notes from a Postnational Geography in, *The Geography of Identity*, edited. P. Yaeger. Ann Arbor: University of Michigan Press, 40-57.

Appadurai, A. 2004. Sovereignty without Territoriality: Notes for a Postnational Geography, in *The Anthropology of Space and Place: Locating Culture*, edited by S.M. Low and D. Lawrence-Zuniga. Massachusetts: Blackwell Publishing, 337-350.

Appadurai, A. 2005. The Production of Locality, in *Counterworks: Managing the Diversity of Knowledge*, edited by R. Fardon. New York: Routledge, 204-225.

Appadurai, A. 2006. *Fear of Small Numbers: An Essay on the Geography of Anger.* Durham NC: Duke University Press.

Appadurai, A. and C. Breckenridge. 1989. On Moving Targets. *Public Culture*, 2, i-iv.

Armbruster, H. 2002. Homes in Crisis- Syrian Orthodox Christians in Turkey and Germany, in *New Approaches to Migration? Transnational Communities and the Transformation of Home*, edited by N. Al-Ali and K. Koser. London: Routledge, 17-33.

Asis, M.M.B. 2002. From the Life Stories of Filipino Women: Personal and Family Agendas in Migration. *Asian and Pacific Migration Journal*, 11(1), 67-94.

Asis, M.M.B. 2003. International Migration and Families in Asia, in *Migration in the Asia Pacific: Population, Settlement and Citizenship Issues*, edited by R.R. Iredal, C. Hawksley, and S. Castles. Massachusetts: Edward Elgar Publishing, 99-120.

Asis, M.M.B. 2006. Living with Migration: Experiences of Left-Behind Children in the Philippines. *Asian Population Studies*, 2(1), 45-67.

Asis, M.M.B., S. Huang, and B.S.A. Yeoh, 2004. When the Light of the Home is Abroad: Unskilled Female Migration and the Filipino Family. *Singapore Journal of Tropical Geography*, 25, 198-215.

Augé, M. 1995. *Non-Places. Introduction to an Anthropology of Supermodernity.* London: Verso.

Auyero, J. 2007. *Routine Politics and Violence in Argentina: The Gray Zone of State Power.* Cambridge: Cambridge University.

Babb, S. 2005. The Social Consequences of Structural Adjustment. *Annual Review of Sociology*, 31, 1-24.

Bailey, A. and Boyle, P. 2004. Untying and Retying Family Migration in the New Europe. *Journal of Ethnic and Migration Studies*, 30(2), 229-241.

Baker, C. and Phongpaichit, P. 2005. *A History of Thailand.* New York: Cambridge University Press.

Bakewell, O. 2008. Keeping Them in Their Place: The Ambivalent Relationship between Development and Migration in Africa. *Third World Quarterly*, 29(7), 1341-1358.

Baldassar, L. 2007. Transnational Families and the Provision of Moral and Emotional Support: The Relationship between Truth and Distance. *Identities: Global Studies in Culture and Power*, 14(4), 385-409.

Barkan, E.R. 2004. America in the Hand, Homeland in the Heart: Transnational and Translocal Immigrant Experiences in the American West, *The Western Historical Quarterly*, 35(3), 331-354.

Barthes, R. 2000. *Camera Lucida: Reflections on Photography*. London: Vintage.

Bayart, J-F. 1979. *L'Etat au Cameroun.* Paris: Presses de la Fondation Nationale des Sciences Politiques.

Beaverstock, J. 2002. Transnational Elites in Global Cities: British Expatriates in Singapore's Financial District. *Geoforum*, 33(4), 525-538.

Bebbington, A. and U. Kothari, 2006. Transnational Development Networks, *Environment and Planning A*, 2006, 38, 849-866.

Becker, E. 1998. *When the War was Over*. New York: Public Affairs.

Bengtson, V.L. and R.A. Harootyan. 1994. *Intergenerational Linkages*. New York: Springer Publishing Company and AARP.

Benwell, M. 2009. Challenging Minority World Privilege: Children's Outdoor Mobilities in Post-Apartheid South Africa. *Mobilities*, 4(1), 77-101.

Bernal, V. 2004. Eritrea Goes Global: Reflections on Nationalism in a Transnational era. *Cultural Anthropology*, 19(1), 3-25.

van Blerk, L. and N. Ansell. 2006. Children's Experiences of Migration: Moving in the Wake of AIDS in Southern Africa, *Environment and Planning D: Society and Space*, 24(3), 449–471.

Blunt, A. 2003. Collective Memory and Productive Nostalgia: Anglo-Indian Homemaking at McCluskieganj. *Environment and Planning D*, 21(6), 717-738.

Blunt, A. 2005. *Domicile and Diaspora: Anglo-Indian Women and the Spatial Politics of Home*. Oxford: Blackwell Publishing.

Blunt, A. 2007. Cultural Geographies of Migration: Mobility, Transnationality and Diaspora. *Progress in Human Geography*, 31(5), 684-694.

Blunt, A. and R. Dowling. 2006. *Home*. Oxon: Routledge.

Blunt, A. and A. Varley. 2004. Introduction: Geographies of Home. *Cultural Geographies*, 11, 3-6.

Blustein, P. 2005. *And the Money Kept Rolling In (and Out): Wall Street, the IMF, and the Bankrupting of Argentina*. New York, NY: Public Affairs Press.

Bourdieu, P. 1990. *The Logic of Practice*. Stanford, CA: Stanford University Press.

Bourdieu, P. 1999. *Acts of Resistance: Against the Tyranny of the Market*. New York, NY: New Press.

Bourdieu, P. 2000. *El sociólogo y las transformaciones recientes de la economía en la sociedad*. Buenos Aires: Libros del Rojas.

Bourdieu, P. 2001. *Las estructuras sociales de la economía*. Buenos Aires: Manantial.

Bourdieu, P. 2002. Habitus, in *Habitus: A Sense of Place*, edited by J. Hillier, and E. Rooksby. Aldershot, Hants: Ashgate, 27-36.

Bourdieu, P. and L.J.D. Wacquant. 1992. *An Invitation to Reflexive Sociology.* Chicago, IL: University of Chicago Press.

Boyer, R. and J.C. Neffa. eds. 2004. *La Economía Argentina y su Crisis (1976-2001): Visiones institucionalistas y regulacionistas.* Buenos Aires: Miño y Dávila Editores.

Bretell, C. 2003. Bringing the City Back: Cities as Contexts for Immigrant Incorporation, in *American Arrivals: Anthropology Engages the New Immigration*, edited by N. Foner. Santa Fe: School of American Research Press, 163-195.

Brickell, K. 2007. Gender Relation in the Khmer 'Home': Post-Conflict Perspectives. London School of Economics: PhD thesis.

Brickell, K. 2008. Tourism-Generated Employment and Intra-Household Inequality in Cambodia, in *Asian Tourism, Growth and Change*, edited by J. Cochrane. Oxford, Elsevier, 299-310.

Brickell, K. Forthcoming. 'We don't forget the old rice pot when we get the new one', Discourses on Ideals and Practices of Women in Contemporary Cambodia. *Signs, Journal of Women in Culture and Society*, 36(2).

Brinkerhoff, J. ed. 2008. *Diasporas and Development: Exploring the Potential* Boulder CO: Lynne Rienner.

Burawoy, M. 2000. *Global Ethnography: Forcers, Connections and Imaginations in a Postmodern World.* Berkeley: University of California Press.

Burrell, K. 2008. Materialising the Border: Spaces of Mobility and Material Culture in Migration from Post-Socialist Poland, *Mobilities*, 3(3), 331-351.

Burton, A. 2005. *African Underclass: Urbanisation, Crime and Colonial Order in Dar es Salaam 1919-1961.* Oxford: James Currey.

Butcher, M. 2009. Ties that Bind: The Strategic Use of Transnational Relationships in Demarcating Identity and Managing Difference, *Journal of Ethnic and Migration Studies*, 35(8), 1353-1371.

Cairns, S. 2004. ed. *Drifting: Architecture and Migrancy.* London, Routledge.

CAPM. 1999. *Corporación Antiguo Puerto Madero, S.A.: Un modelo de gestión urbana, 1989-1999.* Buenos Aires: Ediciones Larivière.

CAPM. 2009. *Puerto Madero: Evolución de Población.* [Online]. Available at: http://www.puertomadero.com/proyecc2.cfm [accessed: 29 August 2009].

Carranza, M. 2005. Poster Child or Victim of Imperialist Globalization? Explaining Argentina's December 2001 Political Crisis and Economic Collapse. *Latin American Perspectives*, 32(6), 65-89.

Cartier, C. 2001. *Globalizing South China.* Oxford: Blackwell.

Cartier, C. 2006. Symbolic City/Regions and Gendered Identity Formation in South China, in *Translocal China: Linkages, Identities, and the Reimagining of Space*, edited by T. Oakes and L. Schein. London: Routledge, 138-154.

Castelain-Meunier, C. 1997. The Paternal Cord. *Reseaux, The French Journal of Communication*, 5(2), 161-176.

Castells, M. 2000. *The Information Age: Economy, Society and Culture*. Oxford: Blackwell.

Castles, S. and M.J. Miller. 2009. *The Age of Migration: International Population Movements in the Modern World*. Basingstoke: Palgrave Macmillan.

Castree, N. 2004. Differential Geographies: Place, Indigenous Rights and 'Local' Resources. *Political Geography*, 23, 133-167.

Centner, R. 2007. Redevelopment from Crisis to Crisis: Urban Fixes to Structural Adjustment in Argentina. *Berkeley Journal of Sociology*, 51, 3-32.

Centner, R. 2008a. *Boom, Bust, and Blur in Buenos Aires: Structurally Adjusted Urbanisms as a Way of Life*. PhD dissertation. Berkeley, CA: Department of Sociology, University of California, Berkeley.

Centner, R. 2008b. Places of Privileged Consumption Practices: Spatial Capital, the Dot-Com Habitus, and San Francisco's Internet Boom. *City & Community*, 7(3), 193-223.

Centner, R. 2009. Conflictive Sustainability Landscapes: The Neoliberal Quagmire of Urban Environmental Planning in Buenos Aires. *Local Environment*, 14(2), 173-192.

Centner, R. Forthcoming. Microcitizenships: Fractious Forms of Urban Belonging after Argentine Neoliberalism. *International Journal of Urban and Regional Research*.

Central Statistical Agency of Ethiopia. 2007. *Population and Housing Survey*. Addis Ababa: The Government of Ethiopia.

Chacko, E. 2003. Ethiopian Ethos and the Making of Ethnic Places in the Washington Metropolitan Area. *Journal of Cultural Geography*, 20(2), 21-42.

Chacko, E. 2008. Washington: From Bi-Racial City to Multi-Ethnic Gateway. In *Migrants to the Metropolis: The Rise of Immigrant Gateway Cities*, edited by M. Price and L. Benton-Short. Syracuse, New York: University of Syracuse Press, 203-226.

Chacko, E. and I. Cheung. 2006. The Formation of a Contemporary Ethnic Enclave: The Case of 'Little Ethiopia' in Los Angeles. In *Race, Ethnicity and Place in a Changing America*. John. S. Frazier and Eugene Tettey-Fio (editors), Global Academic Publishing, pp.131-139.

Chamberlain, M. and S. Leydesdorff. 2004. Transnational Families: Memories and Narratives. *Global Networks*, 4(3), 227-241.

Chau-Pech Ollier, L. and T. Winter. 2006. *Expressions of Cambodia, The Politics of Tradition, Identity and Change*. London: Routledge.

Christensen, P., A. James. and C. Jenks. 2000. Home and Movement: Children Constructing 'Family Time', in *Children's Geographies: Playing, Living, Learning*, edited by S.L. Holloway and G. Valentine. London: Routledge, 139-155.

Christou, A. 2006a. *Narratives of Place, Culture and Identity: Second-Generation Greek-Americans Return 'Home'*. Amsterdam: Amsterdam University Press.

Christou, A. 2006b. Crossing Boundaries – Ethnicizing Employment – Gendering Labor: Gender, Ethnicity and Social Capital in Return Migration. *Social and Cultural Geography*, 7(1), 87-102.

Christou, A. 2006c. American Dreams and European Nightmares: Experiences and Polemics of Second-Generation Greek-American Returning Migrants. *Journal of Ethnic and Migration Studies*, 32(5), 831-45.

Christou, A. 2006d. Deciphering Diaspora – Translating Transnationalism: Family Dynamics, Identity Constructions and the Legacy of 'Home' in Second-Generation Greek-American Return Migration. *Ethnic and Racial Studies*, 29(6), 1040-1056.

Christou, A. and R. King, 2006. Migrants Encounter Migrants in the City: The Changing Context of 'Home' for Second-Generation Greek-American Return Migrants. *International Journal of Urban and Regional Research*, 30(4), 816-35.

Chronopoulos, T. 2006. The *Cartoneros* of Buenos Aires, 2001-2005. *City*, 10(2), 167-182.

Chu, J.Y. 2006. To Be 'Emplaced': Fuzhounese Migration and the Politics of Destination, *Identities*, 13(3), 395-425.

Chu, G.C. and A.W. Schramm. 1991. *Social Impact of Satellite Television in Rural Indonesia.* Singapore: Asian Mass Communication Research and Information Centre.

Ciccolella, P. 1999. Globalización y Dualización en la Región Metropolitana de Buenos Aires. Grandes Inversions y Reestructuración Socioterritorial en los Años Noventa. *Revista Latinoamericana de Estudios Urbano-Regionales*, 25(76), 5-27.

Ciccolella, P. and I. Mignaqui. 2002. Buenos Aires: Sociospatial Impacts of the Development of Global City Functions, in *Global Networks, Linked Cities*, edited by S. Sassen. New York, NY: Routledge, 309-326.

Cieraad, I. 1999. Introduction. Anthropology at Home, in *At Home: An Anthropology of Domestic Space*, edited by I. Cieraad. New York: Syracuse University Press, 1-12.

Clark, W.A.V. and P. Morrison. 1995. Demographic Foundations of Political Empowerment in Multi-Minority Cities, *Demography*, 32(2), 183-201.

Clogg, R. 2002. *A Concise History of Greece*. Cambridge: Cambridge University Press.

Cohen, A.P. ed. 1986. *Symbolising Boundaries: Identity and Diversity in British Cultures*. Manchester: Manchester University Press.

Cohen, P. 2000. From the Other Side of the Tracks: Dual Cities, Third Spaces, and the Urban Uncanny in Contemporary Discourses of 'Race' and Class, in *The Companion to the City*, edited by G. Bridge and S.A. Watson. Oxford: Blackwell, 316-30.

Cohen, R. 2004. Chinese Cockle Pickers, The Transnational Turn and Everyday Cosmopolitanism: Reflections on the New Global Migrants, *Labour, Capital, and Society*, 37, 130-149.

Collins, F.L. 2009. Transnationalism Unbound, Detailing New Subjects, Registers and Spatialities of Cross-Border Lives, *Geography Compass*, 3(1), 434-458.

Conradson, D. and A. Latham. 2005. Transnational Urbanism: Attending to Everyday Practices and Mobilities. *Journal of Ethnic and Migration Studies*, 31(2), 227-233.

Conradson, D. and D. McKay. 2007. Translocal Subjectivities, Mobility, Connection, Emotion, *Mobilities*, 2(2), 167-174.

Cordero-Guzmán, H.R; R.C. Smith, and R. Grosfoguel, eds. 2001. *Migration, Transnationalization, and Race in a Changing New York*. Philadelphia: Temple University Press.

Courtis, C., M.I. Pacecca, D. Lenton, C. Belvedere, S. Caggiano, D. Casaravilla and G. Halpern. 2009. Racism and Discourse: A Portrait of the Argentine Situation, in *Racism and Discourse in Latin America*, edited by T.A. van Dijk. Lanham, MD: Lexington Books, 13-56.

Cresswell, T. 2003. Landscape and the Obliteration of Practice, in *Handbook of Cultural Geography*, edited by K. Anderson et al. London: Sage, 269-281.

Cresswell, T. 2006. *On the Move: Mobility in the Modern Western World*. London: Routledge.

Cupers, K. 2005. Towards a Nomadic Geography: Rethinking Space and Identity for the Potentials of Progressive Politics in the Contemporary City. *International Journal of Urban and Regional Research*, 29(4), 729-39.

Daniels, S. 1993. *Fields of Vision: Landscape Imagery and National Identity in England & the United States*. London: Polity Press.

Datta, A. 2008. Building Differences: Material Geographies of Home(s) among Polish builders in London. *Transactions of the Institute of British Geographers*, 33(4), 518-531.

Datta, A. 2009a. Places of Everyday Cosmopolitanisms: East European Construction Workers in London, *Environment and Planning A*, 41(2), 353-370.

Datta, A. 2009b. 'This is Special Humour': Visual Narratives of Polish Masculinities in London's Building Sites, in *After 2004: Polish Migration to the UK in the 'New' European Union*, edited by K. Burrell. London: Ashgate, 189-210.

Datta, A. and Brickell, K. 2009. 'We have a little bit more finesse as a nation': Constructing the Polish Worker in London's Building Sites, *Antipode: a Radical Journal of Geography*, 41(4), 439-464.

Deleuze, G. and F. Gauttari 1986. *Nomadology: The War Machine*. New York: Semiotext(e).

Deleuze, G. and F. Gauttari 1987. *A Thousand Plateaus*. Minneapolis: University of Minnesota Press.

Deshingkar, P. 2005. *Maximising the Benefits of Internal Migration for Development*, Background Paper for Regional Conference on Migration and Development in Asia, Lanzhou, China, 14-16 March.

Dirección General de Estadística y Censos. 2005. *Anuario Estadístico 2004: Tomo II*. Buenos Aires: Secretaría de Hacienda y Finanzas, Gobierno de la Ciudad de Buenos Aires.

Derks, A. 2005. *Khmer Women on the Move: Migration and Urban Experiences in Cambodia*. Amsterdam: Dutch University Press.

Dobson, M.E. 2009. Unpacking Children in Migration Research. *Children's Geographies*, 7(3), 355-360.

Duval, D.T. 2004. Linking Return Visits and Return Migration Among Commonwealth Eastern Caribbean Migrants in Toronto. *Global Networks*, 4(1), 51-67.

Dze, M. 2005. State Policies, Shifting Cultivation and Indigenous Peoples in Laos. *Indigenous Affairs*, 2(5), 30-37.

Ebihara, M. 1993. Beyond Suffering, The Recent History of a Cambodian Village, in *The Challenge of Reform in Indochina*, edited by B. Ljunggren. Harvard: Harvard Institute for International Development, 149-166.

Edwards, E. 1999. Photographs as Objects of Memory, in *Material Memories*, edited by M. Kwint, C. Breward and J. Aynsberg. Oxford: Berg, 221-236.

Edwards, E. 2002. Material Beings: Objecthood and Ethnographic Photographs. *Visual Studies*, 17(1), 67-75.

Ehrkamp, P. 2006. Rethinking Immigration and Citizenship: New Spaces of Migrant Transnationalism and Belonging, *Environment and Planning A*, 38, 1591-1597.

Elmhirst, R. 2007. Tigers and Gangsters: Masculinities and Feminized Migration in Indonesia. *Population, Space and Place*, 13(3), 225-238.

Evans, M. 2010. Primary Patriotism, Shifting Identity: Hometown Associations in Manyu Division, South West Cameroon. *Africa*, 80(3).

Evans, W.T. 2007. Counter-Hegemony at Work: Resistance, Contradiction and Emergent Culture Inside a Worker-Occupied Hotel. *Berkeley Journal of Sociology*, 51, 33-68.

Faist, T. 2008. Migrants as Transnational Development Agents: An Inquiry into the Newest Round of the Migration-Development Nexus. *Population, Place and Space*, 14, 21-42.

Falzon, M.A. 2003. 'Bombay, Our Cultural Heart': Rethinking the Relation between Homeland and Diaspora, *Ethnic and Racial Studies*, 26(4), 662-683.

Faubion, J. 1993. *Modern Greek Lessons: A Primer in Historical Constructivism*. Princeton: Princeton University Press.

Faulk, K.A. 2008. If They Touch One of Us, They Touch All of Us: Cooperativism as a Counterlogic to Neoliberal Capitalism. *Anthropological Quarterly*, 81(3), 579-614.

Fields, Z. 2008. Efficiency and Equity: The *Empresas Recuperadas* of Argentina. *Latin American Perspectives*, 35(6), 83-92.

Finch, J. and J. Mason. 1990. Divorce, Remarriage and Family Obligations. *Sociological Review*, 38(2), 219-246.

Fox, R.G. 1991. Introduction: Working in the Present, in *Recapturing Anthropology: Working in the Present*, edited by R.G. Fox. Santa Fe, New Mexico: School of American Research Press, 191-210.

Friesen, W., Murphy, L., and Kearns, R., 2005. Spiced-up Sandringham: Indian Transnationalism and New Suburban Spaces in Auckland, New Zealand. *Journal of Ethnic and Migration Studies*, 31(2), 385–401.

Frietag, U. and von Oppen, A. 2009. *Translocality: The Study of Globalising Processes from a Southern Perspective*. Amsterdam: Brill.

Garguin, E. 2007. *'Los Argentinos Descendemos de los Barcos'*: The Racial Articulation of Middle Class Identity in Argentina (1920-1960). *Latin American and Caribbean Ethnic Studies*, 2(2), 161-184.

Geschiere, P. 2009. *The Perils of Belonging: Autochthony, Citizenship and Exclusion in Africa and Europe*. Chicago IL: University of Chicago Press.

Giddens, A. 1990. *The Consequences of Modernity*. Palo Alto: Stanford University Press.

Giddens, A. 1991. *Modernity and Self Identity: Self and Society in the Late Modern Age*, Cambridge, Polity Press.

Gielis, R. 2009. A Global Sense of Migrant Places, Towards a Place Perspective in the Study of Migrant Transnationalism, *Global Networks*, vol. 9(2), 271-287.

Glenn, E.N. 1983. Split Household, Small Producer and Dual Wage Earner: An Analysis of Chinese-American Family Strategies. *Journal of Marriage and the Family*, 45. 35-46.

Glick Schiller, N. And A. Çaglar. (2008) Migrant Incorporation and City Scale: Towards a Theory of Locality in Migration Studies, Willy Brandt Series of Working Papers in International Migration and Ethnic Relations, 2/07, Malmö Institute for Studies of Migration, Diversity and Welfare (MIM) and Department of International Migration and Ethnic Relations (IMER) Malmö University, Malmö: Sweden.

Glick Schiller, N. and G.E. Fouron. 2001. *Georges Woke up Laughing: Long-Distance Nationalism and the Search for Home*. Durham: Duke University Press.

Goldring, L. 1998. The Power of Status in Transnational Social Fields, in *Transnationalism from Below* edited by M.P. Smith and L. E. Guarnizo. New Brunswick, NJ: Transaction Publishers, 165-195.

González Bombal, I. 2002. Sociabilidad en clases medias en descenso: experiencias en el trueque, in *Sociedad y Sociabilidad en la Argentina de los 90*, edited by L. Beccaria et al. Buenos Aires: Editorial Biblos, 97-136.

Gordon, B. 1986. The Souvenir: Messenger of the Extraordinary. *Journal of Popular Culture*, 20(3), 135-146.

Gorman-Murray, A. 2009. Intimate Mobilities: Emotional Embodiment and Queer Migration, *Social & Cultural Geography*, 10(4), 441-460.

Gottesman, E. 2002. *Cambodia After the Khmer Rouge*. New Haven: Yale University Press.

Gough, K. 2008. 'Moving Around': The Social and Spatial Mobility of Youth in Lusaka, *Geografiska Annaler: Series B, Human Geography*, 90(3), 243-255.

Govtrack.us. 2008. A Civic Project to Track Congress. H.R. 2003: Ethiopia Democracy and Accountability Act of 2007. 110th Congress, 2007-2008. http://www.govtrack.us/congress/bill.xpd?bill=h110-2003.

Grimson, A. and G. Kessler. 2005. *On Argentina and the Southern Cone: Neoliberalism and National Imaginaries.* New York, NY: Routledge.

Guarnizo, L. and M.P. Smith. 1998. The Locations of Transnationalism, in *Transnationalism from Below*, edited by M.P. Smith and L. Guarnizo. New Brunswick: Transactions Publishers. 3-34.

Gupta, A. and J. Ferguson 1997. eds. *Culture, Power, and Place.* Durham NC and London: Duke University Press.

de Haas, H. 2006. Migration, Development and Regional Development in Southern Morocco. *Geoforum*, 37(4), 565-580.

Hage, G. 1997. At Home in the Entrails of the West in, *Home/World: Space, Community and Marginality in Sydney's West*, edited by H. Grace., G. Hage., L. Johnson., J. Langsworth and M. Symonds. Sydney: Pluto Press.

Hage, G. 1998. *White Nation: Fantasies of White Supremacy in a Multicultural Society*, Sydney, Pluto Press.

Halfacree, K.H. and P.J. Boyle. 1993. The Challenge Facing Migration Research: The Case for a Biographical Approach. *Progress in Human Geography*, 17(3), 333-348.

Hall, J.M. 2002. *Hellenicity: Between Ethnicity and Culture.* Chicago: University of Chicago Press.

Hammond, L.C. 1999. Examining the Discourse of Repatriation: Towards a More Proactive Theory of Return Migration, in *The End of the Refugee Cycle? Refugee Repatriation & Reconstruction*, edited by R. Black and K. Koser. Oxford: Berghahn Books, 217-244.

Hannam, K., M. Sheller. and J. Urry. 2006. Editorial: Mobilities, Immobilities and Moorings, *Mobilities*, 1(1), 1-22.

Hannerz, U. 1996. Transnational Connections: Culture, People, Places. London: Routledge.

Hannerz, U. 1998. Transnational Research, in *Handbook of Methods in Cultural Anthropology*, edited by H. R. Bernard. London and New Delhi: Altamira Press, 235-256.

Hardill, I. 1998. Gender Perspectives on British Expatriate Work. *Geoforum*, 29(3), 257-268.

Hardt, M. and A. Negri. 2000. *Empire*, Cambridge, MA: Harvard University Press.

Hardy, A. 2004. Internal Transnationalism and the Formation of the Vietnamese Diaspora, in *State/Nation/Transnation*, edited by B.S.A. Yeoh and K. Willis. New York: Routledge, 218-236.

Harney, N. 2007. Transnationalism and Entrepreneurial Migrancy in Naples, Italy, *Journal of Ethnic and Migration Studies*, 33(2), 219-232.

Harper, M. 2005. Introduction, in *Emigrant Homecomings: The Return Movement of Emigrants, 1600-2000*, edited by M. Harper. Manchester: Manchester University Press, 1-14.

Harrison, B. 2004. Snap Happy: Toward a Sociology of "Everyday" Photography, in *Seeing is Believing? Approaches to Visual Research*, edited by C.J. Pole. Oxford: Elsevier, 23-39.

Hatfield, M.E. 2010. Children moving 'home'?: Everyday experiences of return migration in highly skilled households. *Childhood*, 17(2), 243-257.

Helg, A. 1990. Race in Argentina and Cuba, 1880-1930: Theory, Policies, and Popular Reaction, in *The Idea of Race in Latin America, 1870-1940*, edited by R. Graham et al. Austin: University of Texas Press, 37-70.

Henry, L., G. Mohan. and H. Yanacoplous. 2004. Networks as Transnational Agents of Development. *Third World Quarterly*, 25(5), 839–855.

Herzfeld, M. 1986. *Ours Once More: Folklore, Ideology and the Making of Modern Greece*. New York: Pella Publishers.

Hirai, K. 2002. Exhibition of Power- Factory Women's Use of the Housewarming Ceremony in a Northern Thai Village. in *Cultural Crisis and Social Memory-Modernity and Identity in Thailand and Laos*, edited by S. Tanabe and C. Keyes. Honolulu: University of Hawai'i Press, 185-201.

Hirsch, J. 1981. *Family Photographs: Content, Meaning, Effect*. Oxford: Oxford University Press.

Hirsch, M. 1997. *Family Frames: Photography, Narrative, and Postmemory*. London: Harvard University Press.

Ho, E.L.-E. and Hatfield, M.E. 2010. Migration and Everyday Matters: Sociality and Materiality. Population, Space and Place, DOI: 10.1002/psp.636.

Hobsbawm, E. 1991. Introduction. *Social Research*, 58(1), 65-68.

Hochschild, A.R. 1983. *The Managed Heard: Commercialization of Human Feeling*. Berkeley, CA: University of California Press.

Home Office. 2009. *Accession Monitoring Report May 2004-March 2009*, http://www.ind.homeoffice.gov.uk/sitecontent/documents/aboutus/reports/accession_monitoring_report/report-19/may04-mar09?view=Binary, accessed on 11ᵗʰ February 2010.

Hondagneu-Sotelo, P. and E. Avila. 1997. 'I'm here, but I'm there': The Meanings of Latina Transnational Motherhood. *Gender and Society*, 11, 548-571.

Hooper-Greenhill, E. 2000. *Museums and the Interpretation of Visual Culture*. London: Routledge.

Horton, J. 1995. *The Politics of Diversity: Immigration, Resistance, and Change in Monterey Park, California*. Philadelphia: Temple University Press.

Horvath, R.J. 2004, The Particularity of Global Places: Placemaking Practices in Los Angeles and Sydney. *Urban Geography*, 25(2), 92-119.

Hughes, C. 2004. Democracy, Culture and the Politics of Gate-Keeping in Cambodia, The Transnation Goes Home, in *State/Nation/Transnation*, edited by B.S.A. Yeoh and K. Willis. New York: Routledge, 197-217.

Inda, J. and R. Rosaldo. 2002. *The Anthropology of Globalization*. Blackwell, Oxford.

International Organization for Migration (IOM) 2007. *Diaspora Dialogues*. Geneva: IOM Publications.

Iveson, K. 2007. *Publics and the City*, London: Blackwell.

Jackson, K. 1989. *Cambodia, 1975-1978, Rendezvous with Death*. Princeton: Princeton University Press.

Jacobs, J.M. 1996. *Edge of Empire: Postcolonialism and the City*. London and New York: Routledge.

Jacobs, J.M. 2004. Too Many Houses for a Home: Narrating the House in the Chinese Diaspora, in *Drifting: Architecture and Migrancy*, edited by S. Cairns. London: Routledge, 164-183.

Jones, G.A., K. Brickell, S. Chant and S. Thomas-de-Benitez. Forthcoming. *Bringing Youth into Development*. London: Zed Books.

Jusid, J.J. (director). 2002. *Apasionados*. Alquimia Cinema.

Katz, C. 2001. Vagabond Capitalism and the Necessity of Social Reproduction. *Antipode*, 33(4), 709-728.

Kearney, M. 1995. The Local and the Global, The Anthropology of Globalization and Transnationalism, *Annual Review of Anthropology*, 24, 547-565.

Kelly, P. and T. Lusis. 2006. Migration and the Transnational Habitus: Evidence from Canada and Philippines, *Environment and Planning A*, 38, 831-847.

Kenyon, L. 1999. A Home From Home: Students' Transitional Experience of Home, in *Ideal Homes? Social Change and Domestic Life*, edited by T. Chapman and J. Hockey. London: Routledge, 84-95.

Keyes, C.F. 1987. Tribal Peoples and the Nation-State in Mainland Southeast Asia, in *Southeast Asia Tribal Groups and Ethnic Minorities: Prospects for the Eighties and Beyond*. Cultural Survival Report 22, Cultural Survival Inc.

Kiondo, A. 1995. When the State Withdraws: Local Development, Politics and Liberalisation in Tanzania, in *Liberalised Development in Tanzania: Studies on Accumulation Processes and Local Institutions*, edited by P. Gibbon. Uppsala: Nordic Africa Institute, 109-176.

King, A. 2000. Thinking with Bourdieu against Bourdieu: A 'Practical' Critique of the Habitus, *Sociological Theory*, 18(3), 417-433.

King, R. 2000. Generalizations from the History of Return Migration, in *Return Migration: Journey of Hope or Despair?* edited by B. Ghosh. Geneva: International Organization for Migration and the United Nations, 7-55.

King, R. and A. Christou. 2010. Cultural Geographies of Counter-Diasporic Migration: Perspectives from the Study of Second-Generation 'Returnees' to Greece, *Population, Space and Place*, 16(2), 103-119.

King, K.M. and K.B. Newbold. 2008. Return Immigration: The Chronic Migration of Canadian Immigrants, 1991, 1996 and 2001. *Population, Space and Place*, 14(1), 85-100.

King, R. and Skeldon, R. 2010 'Mind the Gap!' Integrating Approaches to Internal and International Migration, *Journal of Ethnic and Migration Studies*, 36(10), 1619-1646.

King, R., R. Skeldon. and J. Vullnetari. 2008. Internal and International Migration: Bridging the Theoretical Divide. *Sussex Centre for Migration Research Working Paper*, 52, 1-49.

Knodel, J. and C. Saengtienchai. 2007. Rural Parents with Urban children: Social and Economic Implications of Migration on the Rural Elderly in Thailand. *Population, Space and Place*, 13, 193-210.

Koser, K. 2007. Refugees, Transnationalism and the State, *Journal of Ethnic and Migration Studies*, 33(2), 233-254.

Koser, K. and R. Black. 1999. The End of the Refugee Cycle?, in *The End of the Refugee Cycle? Refugee Repatriation and Reconstruction*, edited by R. Black and K. Koser. Oxford: Berghahn Books, 2-17.

Kothari, U. 2008. Global Peddlers and Local Networks: Migrant Cosmopolitanisms, *Environment and Planning D: Society and Space*, 26, 500-516.

Lafreniere, B. 2000. *Music Through the Dark, A Tale of Survival in Cambodia.* Honolulu: University of Hawai'i.

Lampert, B. 2009. Diaspora and Development? Nigerian Organizations in London and the Transnational Politics of Belonging. *Global Networks*, 9(2), 162-184.

Langevang, T. and K. Gough. 2009. Surviving through Movement: The Mobility of Urban Youth in Ghana. *Social and Cultural Geography*, 10(7), 741-756.

Latham, A. 2003. Research, Performance, and Doing Human Geography: Some Reflections on the Diary-Photograph, Diary-Interview Method. *Environment and Planning A*, 35(11), 1993-2017.

Laungaramsri, P. 2001. *Redefining Nature: Karen Ecological Knowledge and the Challenge to the Modern Conservation Paradigm.* Chennai: Earthworm Books.

Laungaramsri, P. 2005. Swidden Agriculture in Thailand: Myths, Realities and Challenges. *Indigenous Affairs*, 2, 7-13.

Lavaca. 2004. *Sin patrón: Fábricas y empresas recuperadas por sus trabajadores. Una historia, una guia.* Buenos Aires: Editora Lavaca.

Law, J. 2004. And if the Global were Small and Noncoherent? Method, Complexity and the Baroque. *Environment and Planning D: Society and Space*, 22(1), 13–26.

Law, L. 2001. Home cooking: Filipino Women and Geographies of the Senses in Hong Kong, *Cultural Geographies*, 8, 264-283.

Leontidou, L. 2006. Urban Social Movements: From the 'Right to the City' to Transnational Spatialities and *Flaneur* Activists. *City*, 10(3), 259-68.

Leontidou, L., H. Donnan and A. Afouxenidis. 2005. Exclusion and Difference Along the EU Border: Social and Cultural Markers, Spatialities and Mappings. *International Journal of Urban and Regional Research*, 29(2), 389-407.

Leontis, A. 1997. Whither the Neohellenic? Beyond Hellenicity: Can We Find Another Topos? *Journal of Modern Greek Studies*, 15(2), 217–31.

Lenie, B. 2006. Dutch Moroccan Websites, A Transnational Imagery? *Journal of Ethnic and Migration Studies*, 32(7), 1153-1168.

Levitt, P. 2001. *The Transnational Villagers*. Berkeley, California: University of California Press.

Levitt, P. 2004. Transnational Migrants: When 'Home' Means More than One Country, Migration Information Source, Migration Policy Institute, October 1: http://www.migrationinformation.org/ Feature/print.cfm, accessed on October 15, 2005.

Levitt, P. and S. Khagram 2004. *The Transnational Studies Reader: Intersections and Innovations*. London: Routledge.

Lewinson, A. 2003. Imagining the Metropolis, Globalizing the Nation: Dar es Salaam and National Culture in Tanzanian Cartoons. *City and Society*, 15(1): 9-30.

Ley, D. 2004. Transnational Spaces and Everyday Lives, *Transactions of the Institute of British Geographers*, 29, 151-164.

Ley, D. and A. Kobayashi. 2005. Back to Hong Kong: Return Migration or Transnational Sojourn. *Global Networks*, 5(2), 1470-2266.

Lin, J. 1998. *Reconstructing Chinatown: Ethnic Enclave, Global Change*, Minneapolis, University of Minnesota Press.

Lin, G. and P.A. Rogerson. 1995. Elderly Parents and the Geographic Availability of their Adult Children. *Research on Aging*, 17, 303-331.

Lothar, S. and V. Mazzucato. 2009. Constructing Homes, Building Relationships: Migrant Investments in Houses, *Tijdschrift voor Economische en Sociale Geografie*, 100(5), 662–673.

Low, S.M. and Lawrence-Zuniga, D. 2004. eds. *The Anthropology of Space and Place: Locating Culture*. Massachusetts: Blackwell Publishing.

Lowenthal, D. 1975. Past Time, Present Place: Landscape and Memory. *Geographical Review*, 65(1), 1-36.

Lozada, E.P. 2006. Framing Globalization: Ritual Uses of Photography in Rural China. *Visual Anthropology*, 19(1), 87-103.

Ma 2002. Translocal Spatiality, *International Journal of Cultural Studies*, 5(2), 131-152.

Madison, G. 2006. Existential Migration, *Existential Analysis*, 17(2), 238-60.

Mahler, S. and K. Hansin. 2005. Toward a Transnationalism of the Middle: How Transnational Religious Practices Help Bridge the Divides Between Cuba and Miami, *Latin American Perspectives*, 32(1), 121-146.

Mair, C., Diez-Roux, A. and Galea, S. 2008. Are Neighbourhood Characteristics Associated with Depressive Symptoms? A Review of the Evidence. *Journal of Epidemiology and Community Health*, 62, 940-946.

Magowan, F. 2007. Globalisation and Indigenous Christianity: Translocal Sentiments in Australian Aboriginal Christian Songs. *Identities: Global Studies in Culture and Power*, 14(4), 459-483.

Mandaville, P. 1999. Territory and Translocality: Discrepant Idioms of Political Identity, *Millennium: Journal of International Studies*, 28(3), 653-673.

Marcus, G.E. 1989. Imagining the Whole: Ethnography's Contemporary Efforts to Situate Itself, *Critique of Anthropology*, 9(3), 7-30.

Markowitz, F. and A. Stefansson. eds. 2005. *Homecomings: Unsettling Paths of Return.* Lanham, Maryland: Lexington Books.

Marston, J. 2005. Post-Pol Pot Cambodia, *Critical Asian Studies*, 37(3), 501-516.

Marston, S., J.P. Jones. and K.Woodward. 2005. Human Geography Without Scale. *Transactions of the Institute of British Geographers*, 30(4), 416-432.

Martin, J.N. 1984. The Intercultural Reentry: Conceptualization and Directions for Future Research. *International Journal of Intercultural Relations*, 8(2), 115-134.

Massey, D. 1993. Power-Geometry and a Progressive Sense of Place, in *Mapping the Futures: Local Cultures, Global Change*, edited by J. Bird, B. Curtis, T. Putnam, G. Robertson and L. Tickner. London: Routledge, 59-69.

Massey, D. 1994. A Global Sense of Place, in *Space, Place and Gender* edited by D. Massey. Minneapolis MN: University of Minnesota Press, 146-156.

Massey, D. 1997. A Global Sense of Place, in *Studying Culture: An Introductory Reader* edited by A. Gray and J. McGuigan. London: Arnold.

Massey, D. 1999. Power-Geometries and The Politics of Space and Time, Hettner Lecture 1998, Department of Geography, University of Heidelberg.

Massey, D. 2005. *For Space.* London: Sage.

Mbembe, A. and J. Roitman. 1995. *Figures of the Subject in Times of Crisis. Public Culture*, 7(2), 323-352.

McFarlane, C. 2009. Translocal Assemblages: Space, Power and Social Movements, *Geoforum*, 40, 561-567.

McKay, D. 2006a. Translocal Circulation: Place and Subjectivity in an Extended Filipino Community, *The Asia Pacific Journal of Anthropology*, 7(3), 265-278.

McKay, D. 2006b. Introduction: Finding 'The Field': The Problem of Locality in a Mobile World. *The Asia Pacific Journal of Anthropology*, 7(3), 197-202.

McKendrick, J.H. 2001. Coming of Age: Rethinking the Role of Children in Population Studies. *International Journal of Population Geography*, 7(6), 461-472.

McNeill, D. 2005. Skyscraper Geography. *Progress in Human Geography*, 29(1), 41-55.

McNeill, D. 2006. The Politics of Architecture in Barcelona. *Treballs de la Societat Catalana de Geografia*, 61-62, 167-175.

McNeill, D. 2008. The Hotel and the City. *Progress in Human Geography*, 32(3), 383-398.

McNeill, D. 2009. The Airport Hotel as Business Space. *Geografiska Annaler: Series B, Human Geography*, 91(3), 219-228.

Melamed, D. 2002. *Irse: Cómo y por qué los argentinos se están yendo del país.* Buenos Aires: Editorial Sudamericana.

Mercer, C., B. Page, and M. Evans. 2008. *Development and the African Diaspora: Place and the Politics of Home.* London: Zed.

Mercer, C., B. Page and M. Evans. 2009. Unsettling Connections: Transnational Networks, Development and African Home Associations. *Global Networks*, 9(2), 141-161.

Mignaqui, I. and L. Elguezábal. 1997. Reforma del Estado, políticas urbanas y práctica urbanística. Las intervenciones urbanas recientes en Capital Federal: Entre la 'ciudad global' y la 'ciudad excluyente', in *Postales urbanas del final del milenio: Una construcción de muchos*, edited by H. Herzer. Buenos Aires: Instituto de Investigaciones Gino Germani, 219-240.

Miller, D. 1998. Why Some Things Matter, in *Material Cultures: Why Some Things Matter*, edited by D. Miller. London: UCL Press, 3-21.

Miller, D. 2001. Behind Closed Doors, in *Home Possessions: Material Culture Behind Closed Doors*, edited by D. Miller. Oxford: Berg, 1-19.

Miller, D. 2009. What is a Mobile Phone Relationship? In *Living the Information Society in Asia*, edited by E. Alampay. Singapore: Institute of Southeast Asian Studies, 24-35.

Mills, M-B. 1997. Contesting the Margins of Modernity, Women, Migration, and Consumption in Thailand. *American Ethnologist*, 24(1), 37-61.

Mills, M-B. 2001. Rural-Urban Obfuscations: Thinking about Urban Anthropology and Labor Migration In Thailand. *City and Society*, 13(2), 177-182.

Mitchell, K. 1997. Different Diasporas and the Hype of Hybridity, *Environment and Planning D: Society & Space*. 15(5), 533-553.

Mitchell, K. 2004. *Crossing the Neoliberal Line: Pacific Rim Migration and the Metropolis*, Philadelphia, Temple University Press.

Mizen, P. 2005. A Little 'Light Work'? Children's Images of their Labour. *Visual Studies*, 20(2), 124-139.

Mohan, G., E. Brown, B. Milward and A.B. Zack-Williams. 2000. eds. *Structural Adjustment: Theory, Practice, and Impacts.* London: Routledge.

Molina y Vedia, J. 1999. *Mi Buenos Aires herido: Planes de desarrollo territorial y urbano (1535-2000)*. Buenos Aires: Colihue.

Morgan, N. and A. Pritchard. 2005. On Souvenirs and Metonym: Narratives of Memory, Metaphor and Materiality. *Tourist Studies*, 5(1), 29-53.

Morley, D. 2000. *Home Territories- Media, Mobility and Identity.* London: Routledge.

Muxí, Z. 2004. *La arquitectura de la ciudad global.* Barcelona, Spain: Editorial Gustavo Gili.

Nagel, C. 2002. Reconstructing Space, Re-Creating Memory: Sectarian Politics and Urban Development in Post-War Beirut. *Political Geography* 21(5), 717-725.

Ndjio, B. 2006. Intimate Strangers: Neighbourhood, Autochthony and the Politics of Belonging, in *Crisis and Creativity: Exploring the Wealth of the African Neighbourhood*, edited by P. Konings and D. Foeken. Leiden: Brill, 66-86.

Ndjio, B. 2007. Douala: Inventing Life in an African Necropolis, in *Cities in Contemporary Africa*, edited by Martin Murray and Garth Myers. Basingstoke: Palgrave Macmilllan, 103-118.

Ndjio, B. 2009. Migration, Architecture and the Transformation of the Landscape in the Bamileke Grassfields of West Cameroon. *African Diaspora*, 2(1), 73-100.

Neill, W.J.V. 2005. Berlin Babylon: The Spatiality of Memory and Identity in Recent Planning for the German Capital. *Planning Theory and Practice* 6(3): 335-353.

Nishimoto, Y. 2003. The Religion of the Lahu Nyi (Red Lahu) in Northern Thailand: General Description with Preliminary Remarks. 23, 115-138.

Noble, G. 2005. The Discomfort of Strangers: Racism, Incivility and Ontological Security in a Relaxed and Comfortable Nation. *Journal of Intercultural Studies*, 26, 107-120.

Nyamnjoh, F. 1999. Cameroon: A Country United by Ethnic Ambition and Difference. *African Affairs*, 98(390), 101-118.

Nyamnjoh, F. and M. Rowlands. 1998. Elite Associations and the Politics of Belonging in Cameroon. *Africa*, 68(3), 320-337.

O'Neill, B. and E. Gidengil. 2008. eds. *Gender and Social Capital*. New York, NY: Routledge.

Oakes, T. and L. Schein. 2006. eds. *Translocal China, Linkages, Identities, and the Reimagining of Space*. London: Routledge.

Olson, E. and R. Silvey. 2006. Transnational Geographies: Rescaling Development, Migration, and Religion, *Environment and Planning A*, 2006, 38, 805-808.

Ong, A. 1999. *Flexible Citizenship: The Cultural Logics of Transnationality*. Durham: Duke University Press.

Ong, A. 2003. *Buddha is Hiding*. Berkeley: University of California Press.

Ong, A. 2007. Neoliberalism as a Mobile Technology. *Transactions of the Institute of British Geographers*, 32(1), 3-8.

Page, B. 2007. Slow Going: The Mortuary, Modernity and the Hometown Association in Bali-Nyonga, Cameroon. *Africa*, 77(3), 419-441.

Page, B., M. Evans and C. Mercer. 2010. Revisiting the Politics of Belonging One Decade on in Cameroon. *Africa*, 80(3).

Painter, J. 2000. Pierre Bourdieu, in *Thinking Space*, edited by M. Crang and N. Thrift. London: Routledge, 239-259.

Papastergiadis, N. 2000. *The Turbulence of Migration*. Cambridge: Polity Press.

Parkin, D. 1999. Mementoes as Transitional Objects in Human Displacement. *Journal of Material Culture*, 4(3), 303-320.

Parreñas, R.S. 2001. Mothering from a Distance: Emotions, Gender and Inter-Generational Relations in Filipino Transnational Families. *Feminist Studies*, 27(2), 361-390.

Parreñas, R.S. 2003. The Care Crisis in the Philippines: Children and Transnational Families in the New Global Economy, in *Global Woman: Nannies, Maids, and Sex Workers in the New Economy*. Ehrenreich and Hochschild (eds). New York: Metropolitan Books.

Parreñas, R.S. 2005a. *Children of Global Migration: Transnational Families and Gendered Woes*. Stanford, CA: Stanford University Press.

Parreñas, R.S. 2005b. Long Distance Intimacy: Class, Gender and Intergenerational Relations Between Mothers and Children in Filipino Transnational Families. *Global Networks*, 5(4), 317-336.

Pastor, M. and C. Wise. 1999. Stabilization and Its Discontents: Argentina's Economic Restructuring in the 1990s. *World Development*, 27(3), 477-503.

Pastor, M. and C. Wise. 2001. From Poster Child to Basket Case. *Foreign Affairs*, 80(6), 60-72.

Peleikis, A. 2000. The Emergence of a Translocal Community: The Case of a South Lebanese Village and its Migrant Connections to Ivory Coast. *Cahiers d'études sur la Mediterranée orientale et le monde Turco-Iranien*, 30, 297–317.

Peterson, E.H. 1997. *Subversive Spirituality.* Grand Rapids / Cambridge: WM. B. Eerdmans Publishing Co.

Pieterse, J.N. 2007. *Ethnicities and Global Multiculture: Pants for an Octopus*, London, Rowan and Littlefield.

Pile, S. and M. Keith. 1997. eds. *Geographies of Resistance.* London: Routledge.

Ponchaud, F. 1989. Social Change in the Vortex of Revolution, in *Cambodia, 1975-1978, Rendezvous with Death*, edited by K. Jackson. Princeton: Princeton University Press, pp. 151-177.

Portes, A. and B. Roberts. 2005. The Free Market City: Latin American Urbanization in the Years of the Neoliberal Experiment. *Studies in Comparative International Development*, 40(1), 43-82.

Prévôt Schapira, M.-F. 2000. Segregación, Fragmentación, Secesión: Hacia una Nueva Geografía Social en la Aglomeración de Buenos Aires. *Economía, Sociedad y Territorio*, 2(7), 405-431.

Price, M., I. Cheung, S. Friedman and A. Singer. 2005. The World Settles In: Washington DC as an Immigrant Gateway, *Urban Geography* 26(1), 61-83.

Price, M. and A. Singer. 2008. Edge Gateways: Immigrants, Suburbs, and the Politics of Reception in Metropolitan Washington. In, *Twenty-First Century Gateways: Immigrant Incorporation in Suburban America* edited by A. Singer, S.W. Hardwick and C.B. Bretell. Washington, D.C.: Brookings Institution Press. 138-168.

Raghuram, P. 2009. Which Migration, Which Development? Unsettling the Edifice of Migration and Development. *Population, Space and Place*, 15, 103-117.

Rapport, N. and Dawson, A. 1998. Opening a Debate, in *Migrants of Identity: Perceptions of Home in a World of Movement*, edited by N. Rapport and A. Dawson. Oxford: Berg, 3-38.

Rasmussen, K. 2004. Places for Children – Children's Places. *Childhood*, 11(2), 155-173.

Rath, J. and Kloosterman, R. 2000. Outsiders' Business: A Critical Review of Research on Immigrant Entrepreneurship *International Migration Review*, 34, 657-68.

Rebón, J. 2005. *Desobedeciendo al desempleo: La experiencia de las empresas recuperadas.* Buenos Aires: Ediciones Picaso.

Resurreccion, B. and H.T. Van Khanh. 2007. Able to Come and Go: Reproducing Gender in Female Rural-Urban Migration in the Red River Delta, *Population, Space and Place*, 13(3), 211-224.

Richmond, A.H. 1981. Explaining Return Migration, in *The Politics of Return: International Migration in Europe*, edited by K. Kubat. New York and Rome: Center for Migration Studies, 269-275.

Rigg, J. 2003. *Southeast Asia: The Human Landscape of Modernisation and Development*. London: Routledge.

Rigg, J. 2005. Poverty and Livelihoods after Full-Time Farming: A South-East Asian View, *Asia Pacific Viewpoint*, 46(2), 173-184.

Rigg, J. 2006. Evolving Rural-Urban Relations and Livelihoods in Southeast Asia, in *The Earthscan Reader in Rural-Urban Linkages*, edited by C. Tacoli. London: Earthscan, 68-90.

Rigg, J. 2007. Moving Lives: Migration and Livelihoods in the Lao PDR. *Population, Space and Place*, 13, 163-178.

Robinson, J. 2002. Global and World Cities: A View from off the Map. *International Journal of Urban and Regional Research*, 26(3), 531-554.

Rock, D. 2002. Racking Argentina. *New Left Review*, 17, 54-86.

Rose, G. 2003. Family Photographs and Domestic Spacings, A Case Study. *Trans Inst Br Geogr* 28, 5-18.

Roy, A. 2009. The 21st Century Metropolis: New Geographies of Theory. *Regional Studies*, 43(6), 819-830.

Saïd, E.W. 2003. *Orientalism.* London: Penguin.

Saito, L.T. 1998. *Asian Americans, Latinos and Whites in a Los Angeles Suburb.* Urbana: U. of Illinois Press.

Salih, R. 2001. Moroccan Migrant Women, Transnationalism, Nation-States and Gender. *Journal of Ethnic and Migration Studies*, 27(4), 655-671.

Sandercock, L. 1998. *Towards Cosmopolis: Planning for Multicultural Cities*, Chichester: John Wiley.

Sandercock, L. 2003. Cosmopolis II: Mongrel Cities in the Twenty-first Century, London, Continuum.

Sarlo, B. 2008. Cultural Landscapes: Buenos Aires from Integration to Fracture, in *Other Cities, Other Worlds: Urban Imaginaries in a Globalizing Age*, edited by A. Huyssen. Durham, NC: Duke University Press, 27-49.

Schapendonk, J. 2010. Staying Put in Moving Sands: The Stepwise Migration Process of Sub-saharan African Migrants Heading North in *Respacing Africa* edited by U. Engel and P.Nugent. Leiden: Brill, 113-137.

Schlecker, M. 2005. Going Back a Long Way: 'Home Place', Thrift and Temporal Orientations in Northern Vietnam, *J. Roy. anthrop. Inst.* 11, 509-526.

Schueth, S. and J. O'Loughlin. 2008. Belonging to the World: Cosmopolitanism in Geographic Contexts. *Geoforum*, 39, 926-941.

Selassie, B. 1996. Washington's New African Immigrants. In *Urban Odyssey: Migration to Washington, D.C.*, ed. Frances Carey, 264-275. Washington, D.C.: Smithsonian Institution Press.

Sharp, J.P., Routledge, P., Philo, C. and Paddison R. eds. 2000. *Entanglements of Power: Geographies of Domination / Resistance*. London: Routledge.

Shaw, G and A.M. Williams. 2004. *Tourism and Tourism Spaces*. London: Sage.

Sheller, M. and Urry J. 2006. The New Mobilities Paradigm. *Environment and Planning A*. 38(2), 207-226.

Silvey, R. and V. Lawson (1999) Placing the Migrant. *Annals of the Association of American Geographers*, 89(1), 121-132.

Simmel, G. 2002. The Metropolis and Mental Life. In, *Blackwell City Reader* edited by G. Bridge and S. Watson. London: Blackwell.

Simone, A.M. 2005a. Urban Circulation and the Everyday Politics of African Urban Youth: The Case of Douala, Cameroon. *International Journal of Urban and Regional Research*, 29(3), 516-532.

Simone, A.M. 2005b. Local Navigation in Douala, in *Future City* edited by S. Read, J. Rosemann and J. van Eldijk. London and New York: Spon Press, 212-227.

Sinatti G. 2006. Diasporic Cosmopolitanism and Conservative Translocalism: Narratives of Nation Among Senegalese Migrants in Italy Studies. *Ethnicity and Nationalism*, 6(3), 30-50.

Singer, A. 2004. *The Rise of New Immigrant Gateways*. The Living Cities Census Series. Washington, D.C.: The Brookings Institution.

Skeldon , R. 2003. Interlinkages between Internal and International Migration and Development in the Asian Region, Paper presented at the ad hoc expert group meeting on Migration and Development, ESCAP, Bangkok, 27-29 August.

Skeldon, R. (2008) International Migration as a Tool in Development Policy: A Passing Phase? *Population and Development Review*, 34(1), 1-18.

Skrbis, Z. 1999. *Long-Distance Nationalism: Diasporas, Homelands and Identities*, Avebury: Ashgate.

Slater, D. and Ariztia-Larrain, T. 2009. Assembling Asturias: Scaling Devices and Cultural Leverage, in *Urban Assemblages: How Actor-Network Theory Changes Urban Studies*, edited by I. Farias and T. Bender. London: Routledge.

Smart, A. and G.C.S. Lin. 2007. Local Capitalisms, Local Citizenship and Translocality: Rescaling from Below in the Pearl River Delta Region, China. *International Journal of Urban and Regional Research*, 31(2), 280-302.

Smith, P.M. 1999. Transnationalism and the City. In, *The Urban Moment: Cosmopolitan Essays on the Late 20th Century City* edited by R. Beauregard, E. Body-Gendrot, S. London: Sage Publications. 119–139.

Smith, M.P. 2001. *Transnational Urbanism: Locating Globalization*. Oxford: Blackwell.

Smith, M.P. 2003. Transnationalism, the State, and the Extraterritorial Citizen, *Politics and Society*, 31(4), 467-502.

Smith, M.P. 2005a. Transnational Urbanism Revisited. *Journal of Ethnic and Migration Studies*, 31(2), 235-244.

Smith, M.P. 2005b. Power in Place/Places of Power: Contextualizing Transnational Research, *City & Society*, 17(1), 5-34.

Smith, M.P. 2008. *Citizenship Across Borders: the Political Transnationalism of El Migrante*. Ithaca and London: Cornell University Press.

Smith, M.P. and M. Bakker (2005) The Transnational Politics of the Tomato King: Meaning and Impact, *Global Networks*, 5(2): 129-146.

Smith, M.P. and M. Bakker 2008. *Citizenship Across Borders: The Political Transnationalism of El Migrante*. Ithaca, NY: Cornell University Press.

Smith, M.P. and J. Eade. 2008. Transnational Ties: Cities, Migrations and Identities, in *Transnational Ties: Cities, Migrations and Identities*, edited by M.P. Smith and J. Eade. London: Transaction, 3-13.

Smith, M.P. and L.E. Guarnizo eds. 1998. *Transnationalism from Below*, edited by New Brunswick, N.J.: Transaction Publishers.

Smith, M.P. and L.E. Guarnizo 2009. Global Mobility, Shifting Borders, and Urban Citizenship, *Tijdschrift voor Economische en Sociale Geografie* 100(5), 610-622.

Smith, N. 2008 (1984). *Uneven Development: Nature, Capital and the Production of Space*. Athens, Georgia: University of Georgia Press. 3rd edition.

Smith-Hefner, N. 1999. *Khmer American, Identity and Moral Education in a Diasporic Community.* Berkeley: University of California Press.

Soanes, C. and S. Hawker. 2005. eds. *Compact Oxford English Dictionary.* Oxford: Oxford University Press.

Soja, E. 1989. *Postmodern Geographies: The Reassertion of Space in Critical Social Theory*. London: Verso.

Soja, E. 2000. *Postmetropolis: Critical Studies of Cities and Regions.* London: Blackwell.

Sorkin, M. 1992. ed. *Variations on a Theme Park: New American City and the End of Public Space*. New York: Hill and Wang.

Souter, K. and I. Raja. 2008. Mothering Siblings: Diaspora, Desire and Identity in American Born Confused Desi, *Narrative*, 16(1), 16-28.

Spencer, S., M. Ruhs, B. Anderson, and B. Rogaly, 2007 *Migrants' Lives beyond the Workplace: The Experiences of Central and East Europeans in the UK*. Joseph Rowntree Foundation, www.jrf.org.uk/bookshop.

Sriskandarajah, D. and C. Drew. 2006. *Brits Abroad: Mapping the Scale and Nature of British Emigration*. London: IPPR.

Stewart, K. 1993. *On Longing: Narratives of the Minature, the Gigantic, the Souvenir, the Collection*. Baltimore, Johns Hopkins University Press.

Sturgeon, J.C. 2005. *Border Landscapes: The Politics of Akha Land Use in China and Thailand.* Chiang Mai: Silkworm Books.

Sutton, C.R. 2004. Celebrating Ourselves: The Family Reunion Rituals of African-Caribbean Transnational Families. *Global Networks*, 4(3), 243-257.

Sword, K. 1996, *The Polish Community in Britain*. London: University of London.

Tacoli, C. 2006. ed. *The Earthscan Reader in Rural-Urban Linkages*. London: Earthscan.

Thrift, N. 2004. Intensities of Feeling: Toward a Spatial Politics of Affect. *Geografiska Annaler*, Series B, 57-78.

Thrift, N. 2008. *Non-Representational Theory: Space, Politics, Affect*, London, Routledge.

Tintanji, V., Gwanfogbe, M., Nwana, E., Ndangam, G. and Lima, A. 1988. eds. *An Introduction to the Study of Bali Nyonga: A Tribute to His Royal Highness Galega II, Traditional Ruler of Bali-Nyonga from 1940-1985*. Yaounde: Stardust.

Tolia-Kelly, D. 2004a. Materializing Post-Colonial Geographies, Examining the Textual Landscapes of Migration in the South Asian Home. *Geoforum*, 35(6), 675-688.

Tolia-Kelly, D. 2004b. Locating Processes of Identification: Studying the Precipitates of Re-Memory through Artefacts in the British Asian Home. *Transactions of the Institute of British Geographers*, 29(3), 314-329.

Tolia-Kelly, D. 2008. Motion/Emotion: Picturing Translocal Landscapes in the Nurturing Ecologies Research Project *Mobilities*, 3, 117-140.

Torres, H. 2001. Cambios socioterritoriales en Buenos Aires durante la década de 1990. *Revista Latinoamericana de Estudios Urbano-Regionales*, 27(80), 33-56.

Toyota, M., Yeoh, B.S.A., and Nguyen, L. 2007. Editorial introduction: Bringing the 'Left Behind' Back into View in Asia: A Framework for Understanding The 'Migration-Left Behind Nexus. *Population, Space and Place*, 13, 157-161.

Triandafyllidou, A. and M. Veikou 2002. The Hierarchy of Greekness: Ethnic and National Identity Considerations in Greek Immigration Policy, *Ethnicities* 2(2), 189–208.

Trouillet, M-R. 2003. *Global Transformations: Anthropology and the Modern World*. New York: Palgrave Macmillan.

Tyner, J.A. 2008. *The Killing of Cambodia, Geography, Genocide and the Unmaking of Space*. Aldershot: Ashgate.

UNDP (2009) *Overcoming Barriers: Human Mobility and Development*. Human Development Report, New York: United Nations.

Ureta, S. 2007. Domesticating Homes: Material Transformation and Decoration Among Low-Income Families in Santiago, Chile. *Home Cultures* 4(3), 311-336.

U.S. Census Bureau. 2000. *Census of Population, Summary File 3*. Accessed through American Fact Finder, Washington, D.C.: http://factfinder.census.gov.

U.S. Census Bureau. 2006. *American Community Survey*, Accessed through American Fact Finder, Washington, D.C.: http://factfinder.census.gov.

Um, K. 2006. Refractions of Home, Exile, Memory and Diasporic Longing, in *Expressions of Cambodia, The Politics of Tradition, Identity and Change*, edited by L. Chau-Pech Ollier and T. Winter. London: Routledge, 86-100.

Urry, J. 1990. *The Tourist Gaze: Leisure and Travel in Contemporary Societies*. London: Sage.

Urry, J. 2000. *Sociology Beyond Societies: Mobilities for the Twenty-First Century*, London, Routledge.

Urry, J. 2007. *Mobilities.* Oxford: Polity.

Velayutham, S. and A. Wise. 2005. Moral Economies of a Translocal Village: Obligation and Shame Among South Indian Transnational Migrants. *Global Networks*, 5(1) 27-47.

Vertovec, S. 2006. Is Circular Migration the Way Forward in Global Policy? *Around the Globe*, 3(2), 38-44.

Vertovec, S. and R. Cohen. 2002. *Conceiving Cosmopolitanism: Theory, Context and Practice* Oxford, Oxford University Press.

Vickery, M. 1984. *Cambodia 1975-1982*. Boston: South End Press.

Vieta, M. 2010. The Social Innovations of *Autogestión* in Argentina's Worker-Recuperated Enterprises: Cooperatively Reorganizing Productive Life in Hard Times. *Labor Studies Journal*, 35, 295-321.

Wacquant, L. 1992. Toward a Social Praxeology: The Structure and Logic of Bourdieu's Sociology, in *An Invitation to Reflexive Sociology*, edited by P. Bourdieu and L.J.D. Wacquant. Chicago, IL: University of Chicago Press, 1-59.

Wacquant, L. 2005. Habitus, in *International Encyclopedia of Economic Sociology*, edited by J. Beckert and M. Zafirovski. London: Routledge, 315-319.

Walker, A. 1975. *Farmers in the Hills: Ethnographic Notes on the Upland Peoples of North Thailand.* Pulau Pinang: Phoenix Press.

Walker, A. 1995. eds. '*Mvuh hpa mi hpa'– Creating Heaven, Creating Earth: An Epic Myth of the Lahu People in Yunnan.* Chiang Mai: Silkworm Books.

Walker, A. 2003. *Merit and the Millennium: Routine and Crisis in the Ritual Lives of the Lahu People.* New Delhi: Hindustan Publishing Corporation.

Walsh, K. 2006. British Expatriate Belongings: Mobile Homes and Transnational Homing. *Home Cultures*, 3(2) 119-140.

Walton, J. and C. Ragin. 1990. Global and National Sources of Political Protest: Third World Responses to the Debt Crisis. *American Sociological Review*, 55(6), 876-890.

Warren, S.D. 2009. How Will We Recognize Each Other as Mapuche? Gender and Ethnic Identity Performances in Argentina. *Gender & Society*, 23(6), 768-789.

Werbner, P. 1997. Introduction: The Dialectics of Cultural Hybridity, in *Debating Cultural Hybridity: Multi-Cultural Identities and the Politics of Anti-Racism* edited by P. Werbner, and T. Modood. Atlantic Heights, N.J.: Zed Books, 1-28.

Werbner, P. 1999. Global Pathways. Working-class Cosmopolitans and the Creation of Transnational Ethnic Worlds, *Social Anthropology*, 7(1), 17-35.

Wessendorf, S. 2007. 'Roots Migrants': Transnationalism and 'Return' among Second-Generation Italians in Switzerland. *Journal of Ethnic and Migration Studies*, 33(7), 1083-1102.

Westley, B. 2006. Washington: Nation's largest Ethiopian Community Carves Niche. *USA Today.*

Whitson, R. 2007. Beyond the Crisis: Economic Globalization and Informal Work in Urban Argentina. *Journal of Latin American Geography*, 6(2), 121-136.

Wilding, R. 2007. Transnational Ethnographies and Anthropological Imaginings of Migrancy. *Journal of Ethnic and Migration Studies*, 33(2) 331-348.

Williams, R. 1965. *The Long Revolution*, Hammondsworth, Penguin.

Willis, K., B.S.A. Yeoh. and S.M.A.K. Fakhri. 2002. Introduction: Transnational Elites. *Geoforum*, 33(4), 505-507.

Willis, K.D. and B.S.A. Yeoh. 2000. Gender and Transnational Household Strategies: Singaporean Migration to China. *Regional Studies*, 34(3), 253-264.

Winter, T. 2004. Landscape, Memory and Heritage, New Year Celebrations at Angkor, Cambodia, *Current Issues in Tourism*, 7(4), 330-345.

Wise, A. 2009. Everyday Multiculturalism: Transversal Crossings and Working Class Cosmopolitans, in *Everyday Multiculturalism*, edited by A.Wise and S.Velayutham. London: Palgrave.

Wise, A. 2010. Sensuous Multiculturalism: Emotional Landscapes of Interethnic Living in Australian Suburbia. *Journal of Ethnic & Migration Studies*.

Wise, A. and Velayutham, S. 2009. eds. *Everyday Multiculturalism*, London: Palgrave.

Wise, J.M. 2000. Home: Territory and Identity. *Cultural Studies*, 14(2), 295-310.

Wolbert, B. 2001. The Visual Production of Locality: Turkish Family Pictures, Migration and the Creation of Virtual Neighbourhoods. *Visual Anthropology Review*, 17(1), 21-35.

Wolseth, J. 2008. Safety and Sanctuary: Pentecostalism and Young Gang Violence in Honduras. *Latin American Perspectives*, 35(4), 96-111.

Yenshu, E. 2008. On the Viability of Associational Life in Traditional Society and Home-Based Associations, in *Civil Society and the Search for Development Alternatives in Cameroon*, edited by E. Yenshu. Dakar: CODESRIA, 95-124.

Yeoh, B.S.A., E. Graham and P.J. Boyle. 2002. Migrations and Family Relations in the Asia Pacific Region. *Asian and Pacific Migration Journal*, 11(1), 1-12.

Yeoh, B.S.A., S. Huang, and T. Lam. 2005. Transnationalizing the 'Asian' Family: Imaginaries, Intimacies and Strategic Intents. *Global Networks*, 5(4), 307-315.

Zelinsky, W. and B.A. Lee. 1998. Heterolocalism: An Alternative Model of Sociospatial Behaviour of Immigrant Ethnic Communities. *International Journal of Population Geography*, 4(4), 282-298.

Index

Addis Ababa 13, 16, 163-165, 167, 171-178, 195, 196
Agriculture 44, 50
Affiliation/s, different registers of 3, 7, 8, 11, 13, 15, 16, 17, 27, 129, 133, 134, 136, 138-140, 142, 167, 176, 177, 196
Africa 7, 128-133, 135-137, 139-143, 146, 149, 165, 168, 172, 196
Aid workers 9
Appadurai, Arjun 8, 13, 15, 19, 23, 26, 32, 35, 41, 50, 53, 54, 58, 74, 75, 95, 97, 99, 103, 108, 111, 118, 127, 133, 136, 149, 163, 181, 182, 184
Argentina 17, 109-124, 193
Asylum seekers 9, 13, 130, 166
Athens 13, 17, 145, 149, 150, 152-155, 160, 191
Australasia 7
Australia 15, 93-108, 193, 194

Bali Nyonga 19, 129, 131-135, 137, 139, 140, 143
Belonging 6, 7, 12, 13, 15, 16, 17, 23, 27, 31, 39, 43, 52, 53, 59, 61, 65, 70, 73, 74, 75, 78, 83, 89, 90, 94-97, 99, 100, 102, 106-108, 127, 128, 145-148, 151, 155, 156, 158, 160, 166, 170, 174, 176, 184-186, 189-195, 198
Berlin 13, 145, 149, 150, 159, 191
Bourdieu 11, 12, 76, 99, 112, 114, 117
Buenos Aires 12, 17, 109-124, 187, 192, 193

Cambodia 14, 17, 23-38, 187, 190, 191
Cameroon 16, 17, 19, 23-38, 128-133, 135-143, 196, 197
Capital
    economic 76

material 14, 191
social and cultural 11, 12, 13, 17, 32, 35, 36, 76, 111, 114, 115, 122, 181, 184, 191, 193
symbolic 14, 43, 76
Chinese migrants 93-108, 195
Citizenship 6, 14, 15, 150, 165, 173-175, 182, 185, 193
City/cities 12, 13, 15, 17, 57, 58, 73-79, 81-83, 85, 86-90, 93, 94, 96-98, 105-107, 109-113, 118-120, 122, 123, 124, 130, 131, 134-140, 142, 143, 145-156, 160, 163-174, 176, 177, 181-183, 186, 187, 190-198
Coping strategy/ies 9, 14, 121, 177
Cosmopolitanism/s 6, 14, 17, 41, 87, 105-107, 109, 121, 138, 142, 149, 150, 172, 190, 194

Dar es Salaam 134-136, 196, 197
Diaspora/diasporic 3, 6, 8, 16, 15, 24, 49, 57, 65, 75, 95, 98, 127, 130, 132-134, 137, 140, 141, 143, 145-153, 155, 158, 159, 163-165, 173, 174, 186, 187, 191, 195, 196, 182
Douala 13, 130, 131, 134, 135, 136, 138, 142, 196, 197

Elites 8, 15, 58, 62, 63, 70, 121, 122, 133, 139, 182, 183, 185, 186, 196
Embod-ied/ment 9, 11, 12, 13, 17, 18, 19, 20, 48, 53, 74-76, 90, 95, 97, 99, 108, 109-111, 132, 151, 154, 190, 195
Ethiopia 16-17, 127, 163-177, 186, 195, 196
Europe 7, 77, 79, 82, 130, 131, 142, 149, 150, 152, 174, 192, 193
European Union 11, 14, 73, 78, 82, 158, 175, 193

Everyday life 6, 7, 17, 47, 50, 53, 54, 83, 87, 97, 109, 145, 152, 154, 160, 183, 191, 194

Family 3, 25, 27, 28, 31, 32, 33, 35, 36, 37, 38, 39, 40-54, 58, 60, 61, 63, 64, 65, 66, 68, 69, 84, 97, 105, 129-133, 135, 137, 140, 146, 147, 153-156, 159, 160, 166, 167, 174, 189, 190, 191, 197
Farming 28, 35, 54, 190

Garden/s 15, 17, 34, 58, 66, 68-70, 99
Germany 17, 149, 157, 158
Greece 17, 146, 149, 152-161

Habitus 7, 11, 12, 13, 76, 90, 114, 117, 121, 124
Hannerz, Ulf 8, 15, 58, 136, 143, 181-184, 189
Home/s 5, 6, 7, 10, 11, 12, 14, 15, 16, 17, 18, 19, 23-38, 40, 42, 45-47, 49-52, 55-70, 73-78, 81-83, 85, 90, 93, 95, 98-100, 103, 106, 107, 145-152, 155, 158-161, 166, 170, 173-175, 183-185, 187, 189-198
Homeland 9, 13, 14, 26, 27, 31, 57, 58, 64, 100, 145-150, 152-160, 189, 191, 192, 196
Hometown 16, 31, 84, 87, 90, 99, 128, 129, 133-138, 139, 174, 185
Hometown associations 16, 133-142, 143, 184, 185, 196, 197
Home building/making 5, 14, 19, 55, 56, 59, 62, 63, 64, 66, 70, 96, 100, 132, 133, 187
Hotel 112, 113, 115-118, 121-123, 174, 183, 187, 192, 193, 196

Imaginary/ies 8, 19, 26, 53, 40, 41, 43, 55, 57, 59, 61, 64, 74, 78, 87, 90, 101, 104, 121, 132, 134, 141, 156, 177, 178, 181, 182, 184, 187-192, 195, 198
Indonesia 25
Internal migration 25, 188
Internet 19, 26, 32, 33, 94, 174

Khmer Rouge 23, 27, 28, 29, 32, 34, 36, 37, 52

Lahu 39-54, 190
Lao PDR 25
Latin America 7, 110, 121, 141, 142, 197
Left behind migration 14, 53, 54, 40-44, 46-52, 59, 157, 189, 190, 197
Local-local connections 3-5, 9, 10, 12, 13, 14, 15, 19, 74, 75, 95, 140, 142, 163, 185, 186, 191, 193
Locale/s 3-6, 8, 9, 14, 16, 17, 19, 20, 23, 24, 26, 27, 29, 31, 35, 41, 42, 103, 111, 113, 115, 117, 118, 121-123, 124, 128, 129, 134, 142, 145, 151, 155, 160, 164, 173, 174, 183, 184, 189, 190
Locality/ies 3-5, 7, 8, 10, 14-19, 23, 25-29, 32, 38, 39-41, 50, 52-55, 57-60, 64, 68-70, 73-78, 89, 90, 95-100, 102, 103, 107, 108, 139, 140, 146-151, 155, 163, 164, 168, 169, 170, 176-178, 181-186, 188-191, 195, 196, 198
London 12, 13, 16, 19, 63, 66, 73-90, 122, 127, 129, 130, 132-134, 137, 140, 143, 170, 187, 192, 193, 194, 196, 197
Loyalty/ies 7, 8, 9, 17, 18, 23, 31, 129, 134-137, 139, 140, 142, 143, 182, 196, 197

Material/ity 4, 5, 6, 7, 18, 24, 25, 27, 29, 32, 35, 36, 38, 40, 42, 43, 48, 50, 52, 53, 54, 59, 62, 64, 68, 74-76, 78, 87, 90, 94-97, 99, 100, 103, 104, 107, 108, 112, 113, 129, 132-134, 140, 143, 151, 164, 168, 171, 176, 182, 184, 187, 188, 189, 190, 191, 193, 195, 198
Mexico 9, 17, 182-186
Missionaries 9
Mobile phone 19, 26, 29, 41, 132, 166, 191
Mobilities 14, 17, 18, 41, 56, 77, 81, 83, 86, 87, 111, 150, 152, 163, 164, 172, 177
Movement/s 4-8, 10-19, 26, 39, 42, 57, 58, 66, 69, 73-75, 77, 78, 83, 85-87,

89, 90, 96, 102, 103, 105, 111, 123, 128, 132, 133, 142, 143, 145-149, 166, 168, 172, 174, 177, 185, 186, 193

Multiculturalism 17, 96, 106, 149, 150, 195

Multi-scalar 6, 7, 12, 13, 17, 25, 123, 155, 172, 185-188, 195, 196, 198

Multi-sited 7, 13, 17, 24, 38, 145, 148, 176, 178, 183, 184

Neighbourhood/s 5, 7, 9, 11, 12, 13, 15, 16, 17, 58, 73-78, 82-90, 94-100, 103-106, 108, 110-115, 117, 118, 123, 150, 163-172, 175-177, 181, 182, 186, 187, 190, 192-197

New York 13, 145, 149, 150, 191

North America 7, 174

Nostalgia 5, 24, 44, 50, 52, 54, 64, 100, 104-106, 108, 122, 141, 190, 198

Otherness 5, 14, 16, 18, 73, 191

Photograph/y 7, 15, 25, 27, 28, 29, 30, 31, 32, 33, 35, 36, 61, 70, 77, 80, 166, 189

Poland 17, 18, 19, 73, 77-79, 83-87, 89, 90, 193

Polish migrants 12, 18, 19, 73-90, 93, 192-194

Politics 4, 6-9, 11, 13-18, 20, 24, 25, 27, 33, 39, 43, 44, 56, 73, 74, 78, 82, 87, 90, 94, 97, 148-151, 156, 165-168, 171-173, 175-177, 181-198

Puerto Madero 110-119, 121-124

Refugee/s 8, 9, 10, 13, 57, 166, 188, 192

Regional 4, 10, 11, 12, 14, 78, 133, 138, 156, 166, 168, 170, 173, 177, 182, 185, 193, 197

Restaurant/s 17, 18, 35, 85, 93-95, 98, 147, 166-169, 171, 194, 195

Return migration 55-57, 59, 60, 62, 66, 69, 79, 80, 83-85, 121, 129, 132, 137, 138, 141, 152-161, 164, 166, 174, 175, 186-193, 197, 198

Rural-urban migration 4, 10, 12, 14, 24, 25, 26, 27, 28, 38, 42, 43, 46, 190, 197

Siem Reap 23-38, 190, 191

Signage 96-105, 107, 108, 195

Singapore 12, 15, 17, 18, 55-70, 187, 189

Situatedness 6, 7, 8, 9, 10, 11, 15, 20, 25, 54, 74, 76, 90, 145, 147, 155, 181, 183, 188

SMS 19, 43, 46

Southeast Asia 7, 25, 35, 43, 68

Sydney 13, 15, 93-108, 181, 183, 188

Tanzania 16, 17, 128, 133, 135-137, 139-142, 196, 197

Temporality/ies 16, 18, 38, 48, 50, 64, 82, 136, 146, 148, 153, 155

Thailand 14, 17, 25, 36, 39-54, 190

Tourism 12, 25, 28, 29, 35, 36, 38, 44, 117, 190, 191

Translocalism 70, 127-129, 132, 133, 136-143, 147

Translocality 3-10, 13-19, 24, 25-29, 31, 32, 38, 41, 42, 53, 54, 55-58, 60, 61, 62, 65, 66, 69, 70, 74, 77, 95, 108, 111, 113, 118, 122-124, 145-147, 150-152, 155, 163-165, 172, 174-177, 181-197

Translocal
    city 74-77, 86, 89, 90, 193
    village 40, 41, 42, 44, 46, 49, 50, 53, 54

Transnationalism 3, 4, 7-10, 24, 25, 27, 41, 42, 56, 133, 136, 137, 140-143, 147, 148, 181-190, 192-197
    from below 10
    from between 74
    Grounded 3, 4, 6, 7, 8, 9, 10
    Internal 25

United Kingdom 17, 56, 60, 61-68, 70, 73-90, 129, 133-135, 139, 140, 142, 189, 193, 196, 197

United States of America 13, 17, 116, 149, 163-170, 173-177

Urban 3-6, 10-12, 14-17, 19, 24-29, 34-36, 38, 39, 44, 46, 48, 49, 58, 70, 73-78, 82, 87, 90, 93, 94, 96, 105, 107, 110, 112, 114, 122, 123, 127-143, 145-152, 160, 164, 165, 167-169,

171-174, 177, 181, 182, 185, 187,　　Washington D.C. 13, 16, 163-178, 186,
188, 190, 191, 193-197　　　　　　　195

Vietnam 25, 28
Visual narrative/s 18, 77, 89